D1214879

HVAC Control In the New Millennium

HVAC Control In the New Millennium

by
Michael F. Hordeski

Published by
THE FAIRMONT PRESS, INC.
700 Indian Trail
Lilburn, GA 30047

Library of Congress Cataloging-in-Publication Data

Hordeski, Michael F.
HVAC control in the new millennium / by Michael F. Hordeski
p. cm.
Includes bibliographical references and index.
ISBN 0-88173-339-3
Heating—Control. 2. Air Conditioning—Control. 3. Ventilation—Control. I. Title.

TH7466.5 .H68 2001
697—dc21

00-051415

Published by The Fairmont Press, Inc.
700 Indian Trail
Lilburn, GA 30047

Printed in the United States of America

10 9 8 7 6 5 4 3 2 1

Distributed by Marcel Dekker, Inc.

Headquarters:
270 Madison Avenue, New York, New York 10016
Telephone: 212-696-9000; Fax: 212-685-4540

Distribution and Customer Service:
Cimarron Road, Monticello, New York 12701
Telephone: 845-796-1919; Fax: 845-796-1772

Eastern Hemisphere Distribution:
Hutgasse 4, Postfach 812, CH-4001, Basel, Switzerland
Telephone: 41-61-260-6300; Fax: 41-61-260-6333

Table of Contents

The use of wireless LANs may double or even triple over the next few years. Wireless LANs allow maintenance workers to report problems using wireless scanners, portable computers and other hand-held devices.

Improved efficiency is possible by applying control concepts such as ratio and feed-forward control. Further HVAC system improvement and optimization may be accomplished with advanced modeling and simulation techniques which are used to develop alternate solutions and allow comparative analysis.

The computer evolution is on an upward spiral, which concerns the nature of intelligence, its significance and its ultimate realization in machines that will cause the world to be transformed.

Future HVAC systems will involve mixing intelligent and non-intelligent devices. Intelligent systems technology, such as fuzzy logic, expert systems and neural networks will find practical applications to provide intelligence for control management and optimization. Fuzzy logic will be used in supervisory controllers based from operator experience and experimental results rather than mathematical models. While traditional control and the computer code used to implement it relies heavily on absolutes like true/false, yes/no or on/off conditions, fuzzy logic employs a more human way of thinking about a problem. This book is about a future which involves a transformation of living and working habits at many levels.

HVAC systems may be integrated into plant-wide supervisory control and data acquisition (SCADA). In the new age of component-based software, objects may be used to snap together control systems using flexible open standards based on general PC technologies. Hardware will also evolve towards more plug-in components with standard interfaces for both hardware and software compatibility.

Control technology continues to be driven by information and the means of delivering it. The chief information enabler is software. Software areas of special interest include operating systems, open systems, object-based technologies, and Internet communications. This book looks at the key trends in these areas, and discusses the impact each is likely to have on control systems.

Many HVAC systems will use Windows NT and open standard networks based on BACnet/TCP/IP Ethernet with open data exchange standards. These systems will use standard Web technology for monitoring and control. Using a standard Web browser, alarm and event windows are implemented as Java applets that are automatically downloaded to the browser.

Control software will include object-oriented languages, with fill-in-the-blank templates for control loops. PCs will be used to configure and test the control package off-line and then download it into the real-time system. The WIMP (windows, icons, mouse, pointers) concept will take over the role of typewriter-like keyboards. Multi-language control software will reduce the chances of operator error by displaying information in German, French and other languages.

The technology of heating has gone from simple attempts to keep space areas warm to sophisticated computerized systems that track occupants levels, seasons, time-of-day, outdoor temperature and humidity conditions. Ventilation will continue to play a key role in occupant comfort levels and plant efficiency. Air cleanliness will become more important and more emphasis will be placed on dust, bacteria, odors and toxic gases which may affect human health and comfort.

The close control of the amount of outdoor air used in the HVAC system is only one of the areas being mined for energy savings. Excessive energy consumption is often due to the infiltration of outdoor air which can account for about 1/2 of the total heating and cooling loads in some public buildings. Building conditioning accounts for about 20% of the total energy consumed in the U.S., so computer-optimized HVAC systems can make a major contribution in reducing our national energy use in the next millennium.

The nature of utility demand-side management (DSM) is changing. Many energy efficiency measures are difficult to promote through the utilities. Equipment efficiency standards are expected to lower national electricity use by several percentage points.

The computer revolution is moving towards an emancipation with an overwhelming and comprehensive impact. When the machine age began, it surged into motion and almost ran its course before most people were aware of what had happened. We have the gift of better communications which help us to contemplate some of the amazing changes that may take place. This book attempts to provide an understanding of the nature of these changes. It uses case histories to allow a comparative analysis for current and future HVAC systems. It compares intelligent and non-intelligent devices and assesses the capabilities and limitations of intelligent devices and their communications.

This book shows how to differentiate between the available communication technologies, predict the value of improvements gained by intelligent field devices and grasp the business drivers behind integration, industrial web technologies for HVAC and security features of industrial networks.

Chapter 1 introduces heat, ventilation and damper control trends as well as air conditioning trends. Basic concepts such as the properties of air, filtering and heat transfer are described. Zone control and year-around air conditioning are explained along with air washing. The concept of HVAC intelligence is introduced and advanced motion control for dampers including feedback devices, networking and PCs with digital drives are covered.

Chapter 2 considers energy and power management trends. Distributed control levels include HVAC, communications and information management. Topics include direct digital control and proportional zone control.

Future computer-based systems for intelligent buildings and controls may use biochemical computers. Device communications for the intelligent I/O appliances used in complex adaptive systems may take place in industrial networks that use wireless connections.

Control technology and microelectronics are the main subjects of Chapter 3. After a summary of control modes and their characteristics, supervisory control techniques are considered for system optimization and old and new control concepts are compared. Advances in microelectronics push much of the new computer applications. The computer evolution is due to shrinking devices. Nanotechnology will produce molecular and protein machines which are new types of molecular computers. Universal assemblers and disassemblers will build and rebuild future products. Reconfigurable computers with reconfigurable hardware are even closer in our future.

Chapter 4 looks at advanced HVAC control, information technology and open systems trends. Semiconductor manufacturing involves advanced HVAC control concepts including air contamination, zone optimization, pressure control, flow control and pollution control. Intelligent alarms will use expert systems reasoning. Fault tolerant systems can use passive redundancy or active redundancy.

Windows NT has become a de facto standard for control systems while Ethernet use has expanded as a plant control network. If feed-forward is not used properly for HVAC, a number of errors can be propagated. This is explained in Chapter 4.

Object technology is growing as it fuels open protocols and open control devices. This trend is producing open building controls.

Chapter 5 is concentrated on PC-based control, software and bus trends. Object concepts are explained along with trends in operating systems including Windows CE, Windows NT-2000, UNIX and Linux. ActiveX and Java are important tools in PC-based control, but the Internet remains as the key to the future. Future computers will use natural languages, chatterbots and voice recognition software. Search engines will become smarter as the web advances into the future. Smart software will use Bayesian Networks, personal slaves and wizards.

Processing power depends on fabrication technology which has physical limitations. Chip multiprocessors are expected to proliferate in the near future along with faster bus connections and products like wearable electronic displays.

Chapter 6 is concerned with artificial intelligence, fuzzy logic and control. This includes the different levels and areas of artificial intelligence including speech recognition systems, expert systems and neural networks. Fuzzy control can be used for temperature balancing or the overshoot that occurs during heat-up. Fuzzy control has been used to reduce the variability in product properties by 30% to 40%. Neural networks are being used in burner management systems for commercial power plants.

Chapter 7 is concerned with computer networks and security. The International Standards Organization model provides the functional layers for networking. Ethernet trends include wireless Ethernet. Internet-enabled networks require security techniques such as digital certificates and passwords.

Chapter 8 covers open systems and device networks. Building automation trends include the use of BACnet with its conformance classifications and interoperable functions. LonWorks control is an alternative and BACnet and LonMark may be used together. Other industrial networks include DeviceNet, CAN, ControlNet, Seriplex, AS-i, HART, Foundation Fieldbus, Profibus, CAB and MODBUS. Some of these networks require devices with healthy processing which makes them more expensive than the control devices used in a typical commercial or institutional building for environmental control.

Chapter 9 concentrates on building automation, wireless technology and the Internet. A variety of BACnet projects have been completed or are underway. There is a merging of information and control in building automation. This will continue as use of the Internet and wireless technology grow. The IrDA technology for infrared control may become prevalent. Cellular problems include security. The use of wireless LANs will grow along with the growth of pocket communications. Embedded Web pages may be included in many controllers for troubleshooting and maintenance. Personal digital assistants (PDAs), pagers and smart cell phones will share Windows data.

Smart antennas use their intelligence to sense the radio environment and optimize their performance. As Internet connections increase and the Internet grows, there will be new public networks that will reduce Internet bottlenecks. These new public networks will have added intelligence for configuration chores, managed backbones and security.

Many thanks to Dee who kept this project moving along on schedule.

Chapter 1

Heat, Ventilation and Damper Control Trends

The first attempts at environmental control involved heat and ventilation. The control of heat goes back to prehistoric times and probably first developed in colder climates where warmer temperatures were needed for both comfort and health. The technology of heating has advanced from simple attempts to keep the body warm to the sophisticated systems of maintaining stabilized environments in order to minimize the heat loss from the body or the structural surfaces of the room.

Ventilation also goes back to very early periods in history. The use of slaves to wave large fans and fan-like devices was an attempt to improve the ventilation. Rooms or buildings were also arranged to take advantage of the rising and setting of the sun as well as prevailing breezes and winds. By the end of the 18th century significant advances were being made in heating and ventilating.

In the early stages of the Industrial Revolution, ventilating and heating acquired increased importance since it affected the work efficiency and the health of the workers. This resulted in the creation of ventilation systems to remove contaminants from the air.

Air Conditioning

Air conditioning is a more recent development and encompasses all aspects of environmental control. In addition to the control of temperature, both humidity (the moisture content of the air) and air cleanliness are regulated by air conditioning.

The earliest attempts at air conditioning involved the placing of wet cloths over air passages such as window openings and doors to cool the air. Air conditioning technology began to progress much further when in the 1840s, several systems evolved for both cooling and humidifying rooms.

The first units were developed by textile manufacturers to reduce static electricity in the air. Later, adaptations were made by other industries.

Although developments in air conditioning technology increased rapidly in the first part of the 20th century, the widespread use of air conditioning in buildings did not take place until the post-World War II period.

Unlike the early forms of air conditioning which were designed to cool the air and add moisture to it, modern air conditioning systems can control temperature, air moisture content, air cleanliness, and air movement. The modern systems truly condition the air, rather than simply cool it.

Heating Systems

Many different methods have been devised for heating buildings. Each has its own characteristics and most of them have at least one undesirable characteristic such as the cost of fuel, expensive to install or inefficient heating. Most of these heating methods can be classified according to one of the following four criteria:

- heat conveying medium,
- fuel used,
- nature of the heat, and
- efficiency of the method.

The heat conveying medium is the substance or combination of substances that carries the heat from its point of origin to the area being heated. The four basic mediums for conveying heat are air, water, steam and electricity.

Throughout history different types of wood, coal, oil, and gas products have been used as fuels for producing heat. Electricity can be considered both as a fuel and a heat conveying medium. Each type of heating fuel has its characteristics and the advantages of one over another depends on variables such as availability, efficiency and cost.

Heating Methods

Heating methods can also be classified with respect to the nature of the heat applied. For example, the heat may be of exhausted steam or even consist of exhaust gases from waste combustion.

The different heating methods vary considerably in efficiency and desirability. This is due to different but often interrelated factors such as fuel cost, medium employed and the type of heating unit. The integration of these interrelated components into a single operating unit results in the heating system.

Among the variety of heating systems available are the following broad classifications:

- warm-air heating systems,
- hot-water heating systems,
- steam heating systems, and
- electric heating systems.

These classifications of heating systems are based on the heat conveying method used.

Ventilating Systems

Ventilating is closely interrelated with heating in its various applications. One ventilating system may be used to provide fresh air while another must remove noxious gases or other dangerous contaminants from the enclosure. The type and design of ventilating system employed will depend upon a number of different factors, including:

- building use or ventilating purpose,
- size of building,
- geographical location, and
- heating system used.

Buildings located in the tropical or semi-tropical zones present different ventilating problems from those found in the temperate zones. The differences are so great that they often result in different architectural forms especially before the widespread use of air conditioning.

The early southern houses were constructed with high ceilings, large porches which sheltered sections of the house from the direct rays of the sun and large window areas for ventilation. They were also usually situated on the site so that halls, major doors and sleeping areas faced the direction of the prevailing winds. Today, with air conditioning, these considerations have not been given as much importance.

Ventilation Requirements

Ventilation is the process of supplying or removing air to or from a building or space. The air may or may not have been conditioned. Methods of supplying or removing the air include natural ventilation and mechanical methods.

Natural ventilation involves opening or closing wall vents and by using roof vents or ventilators. Mechanical ventilation is provided by the use of fans or other means that force the air through the space to be ventilated.

Air Conditioning Trends

There has been an increasing use of year-round air conditioning systems that provide heating, ventilating, and cooling. These systems condition the air by controlling its temperature (warming or cooling), cleanliness, moisture content and movement. Thus, the true meaning of the term air conditioning has been achieved.

Air conditioning can be treated as the simultaneous control of those factors that affect the conditions of the atmosphere in a structure. These factors can include temperature, humidity, air motion and distribution, dust, bacteria, odors, and toxic gases which affect human health and comfort.

Air that has been properly conditioned generally has a combination of processes performed on it. It may have been heated or cooled and may have had moisture removed from it (dehumidified). Fans may be used to move the air and filters may be used for cleaning. The air acts as a vehicle to transport heat and moisture from one area. Air movement is one important technique of energy conservation.

Properties of Air

Air is a mixture of two main gases, approximately 23 parts of oxygen and 77 parts nitrogen by weight. The other gases in air includes carbon dioxide, carbon monoxide, ozone and neon in small quantities.

Carbon monoxide is not generally present in the atmosphere, but is found in congested motor traffic. It is dangerous and is also given off by stoves and furnaces. Air containing carbon monoxide in a concentration greater than 0.10% is fatal to humans.

Oxygen is the most important constituent of air, it makes up about 1/4 of the air by weight and 1/5 of the air by volume and provides the existence of all animal life. Nitrogen is an inert gas which acts to dilute oxygen.

Air Circulation

Air circulation is important for two reasons. Air that is moving can carry away the moisture and heat surrounding the body. If this is not done, the occupants can be uncomfortable even when the relative humidity of the area is low.

Air that is drawn into the conditioner and passed out over the evaporator will remove the moisture from the room which is condensed and passed through the drain.

Although the movement of air is essential, direct drafts should be avoided. Condensation on the evaporator surface during summer operation produces a certain amount of cleaning since the moisture absorbs impurities from the air as it passes over the evaporator surfaces. The condensation of the moisture continues to carry off these impurities to the drain.

Filters

Air cleaning and filtering devices can eliminate most particles carried in the air that are detrimental to health and comfort. These include dust, fumes, and smoke. Dust and fumes tend to settle in still air, but smoke is actuated by motion and if not removed, will remain in motion in the air.

Air washing can be effective in removing dust and those fumes and smoke that are soluble in water, but carbons, soot, and similar substances cannot be removed by this type of cleaning. It is possible to cleanse the air of these by substances using either the dry and viscous filters that have been developed.

Heat Transfer

When bodies of unequal temperatures are placed near each other, heat leaves the hotter body and is absorbed by the colder body until the temperature of each is equal. The rate at which the heat is absorbed by the colder body is proportional to the difference of temperature between the two bodies, the greater the difference of temperature the greater the rate of flow of the heat.

Heat is an active form of energy, much the same as mechanical and electrical energy. Heat can be transferred by three methods, conduction, convection, and radiation.

Conduction and Convection

Conduction refers to heat transfer that takes place mainly in solids where the heat is passed from one molecule to another without any noticeable movement of the molecules.

Conduction is the transfer of heat through a substance. Conductivity may be defined as the relative value of a material as compared with a standard, which affords a passage for heat. A poor conductor is known as a nonconductor or insulator. Copper is a good conductor, while the various materials used to insulate buildings are poor conductors. Any substance, which is a good conductor of electricity is also a good conductor of heat.

Convection is a heat transfer that takes place in liquids and gases where the molecules carry the heat from one point to another. Convection involves the transfer of heat due to the motion of the heated matter. Since motion is a requirement for convection, it can only take place in liquids and gases.

Radiation

Radiation is heat transfer in a wave motion, such as light or radio waves. It takes place through a transparent medium without affecting the medium's temperature. An example of this is solar radiation.

Radiation, as a heat loss, involves the throwing out of heat in rays. The heat rays proceed outwards in straight lines and the intensity of the heat radiated from any one source becomes less as the distance from the source increases.

The amount of heat loss through radiation depends upon the temperature of the surrounding surfaces. The colder these surfaces are, the faster and greater will be the heat loss from a heated object. If the surfaces are warmer than the heated object, then heat will be radiated from these surfaces to the object.

Radiation, conduction, and convection are often interrelated. Heat from a burning fuel passes to the metal of the heating surface by radiation, through the metal by conduction, and is transferred to water by convection. The circulation needed for convection is caused by the variation in the weight of the water due to temperature differences. The water next to the heating surface receives heat and expands becoming lighter. It rises displacing the colder and heavier water above.

Mean Radiant Temperature

The mean radiant temperature of the surfaces of an enclosure is important when dealing with heat loss by radiation. The mean radiant temperature (MRT) is the weighted average temperature of the surrounding surfaces.

The significance of the mean radiant temperature becomes apparent when compared with the clothed body of an adult at 80°F or 26.7°C. If the MRT is below 80°F, the human body will lose heat by radiation to the surfaces of the enclosure. If the MRT is higher than 80°F, the opposite effect will occur.

Sensible Heat

Sensible heat is a form of heat that causes a change in the temperature of a substance and can be measured by a dry-bulb thermometer. There is no change in moisture content. It is referred to as sensible heat because it can be sensed by instruments or the touch. When the temperature of water is raised from 32 to 212°F, an increase in the sensible heat takes place since these are dry-bulb measurements.

Specific Heat

This refers to the amount of Btus required to raise the temperature of one pound of a substance one degree Fahrenheit. The specific heat of a substance is the ratio of the quantity of heat required to raise its temperature one degree Fahrenheit to the amount required to raise the temperature of the same weight of water one degree Fahrenheit. This can be expressed by the following formula:

$$\text{Specific heat} = \frac{\text{Btu to raise temperature of substance } 1^\circ\text{F}}{\text{Btu to raise temperature of same weight water } 1^\circ\text{F}}$$

The standard is water at 62°F to 63°F which has a rating of 1.00 on the specific heat scale. Specific heat represents the Btu's required to raise the temperature of 1 pound of a substance 1°F.

The specific heat of water is defined as 1.00, but the amount of heat required to raise the temperature of different substances through some temperature range will differ. Water has a large heat capacity and it is used as a standard. Since 1 pound of water requires one Btu to raise its temperature one degree Fahrenheit, it is rated on the specific heat scale at 1.00. Iron has a lower specific heat with an average rating of 0.130, ice is 0.504 and air is 0.238. The more water a substance contains, the higher the specific heat.

Latent Heat

Latent heat is that quantity of heat which disappears or becomes concealed in a body while producing some change in it other than a rise of temperature.

Latent heat means hidden heat and it is a form of heat that causes a substance to change its physical state from a solid to liquid, a liquid to vapor, or the reverse. When a liquid is evaporated to a gas, the change of physical state is always accompanied by the absorption of heat. Evaporation provides a cooling effect on the surroundings of the liquid since the liquid must obtain from its surroundings the required heat to change its molecular structure. This type of action takes place in the evaporator of an air conditioner. All liquids tend to saturate the surrounding space with its vapor. This property is important in air conditioning.

Latent Heat of Fusion

The change of a substance from a solid to liquid or from a liquid to a solid involves the latent heat of fusion. One pound of water at a temperature of 32°F requires the extraction of 144 Btu to cause it to freeze into solid ice at 32°F. All solid substances have a latent-heat value and the amount required to convert it, or affect a change of state, is called the latent heat of fusion. This heat cannot be measured with a thermometer since the heat is absorbed or expanded in the change of state process.

Latent Heat of Evaporation

The change of a substance from a liquid to a vapor or from a vapor to a liquid involves the latent heat of evaporation. The conversion of 1 pound of pure water at 212°F to steam at 212°F requires 970 Btu at the pressure of the atmosphere found at sea level. After the 970 Btus have been used, the further addition of heat serves only to heat the steam.

Superheat

Superheat is the sensible heat that is absorbed by a vapor or gas not in contact with its liquid. It does not follow the temperature-pressure relationship and refers to the sensible heat absorbed by the vapor raising the temperature of the vapor or gas without any appreciable change in pressure.

A gas is usually considered as a vapor in a highly superheated state or as a vapor not near its condensing point. Water in the air that is close to the condensing point is water vapor. Since as superheat is sensible heat, its effect can be measured with a thermometer and as a temperature rise. A 10°F superheat means a vapor has absorbed enough heat to raise the vapor temperature 10°F above the temperature of vaporizing liquid.

When a gas is condensed to a liquid, the change of physical state is always accompanied by the giving up of heat. This takes place in the condensing unit of an air conditioner due to the mechanical work exerted on the gas by the compressor.

Heat Energy

Heat is a form of energy that is in transit between its source and destination point. Heat energy only exists while flowing between the source and destination.

Heat energy is similar to work energy, which is another form of energy in transit. The main difference is that heat energy is energy in transit as a result of temperature differences.

British Thermal Unit

The British thermal unit is a unit of heat or heat energy. Since 1929 British thermal units have been defined as one Btu being equal to 251.996 IT (International Steam Table) calories, or 778.26 feet-pound of mechanical energy units (work). Since one IT calorie equals 1/860 of a watt-hour, one Btu is equivalent to about 1/3 watt-hour.

Prior to 1929, one Btu was defined as the amount of heat necessary to raise 1 pound of water 1°F. Because of some difficulty in determining the exact value of a Btu, it was redefined in terms of a more fundamental electrical unit.

Heat and Work

Energy can be defined as the ability to do work or move against a resistance. Work can be defined as the overcoming of resistance through a certain distance using the expenditure of energy.

Work is measured by a standard unit called the foot pound. This is defined as the amount of work done in raising 1 pound the distance of 1 foot, or in overcoming a pressure of 1 pound through a distance of 1 foot.

A relationship between work and heat exists and is referred to as the mechanical equivalent of heat. This states that one unit of heat is equal to 778.26 foot-pounds of work.

The relationship of the mechanical equivalent of heat was first established by experiments conducted in 1843 by Dr. Joule of Manchester, England. He determined by experiment that when 772 foot-pounds of energy had been expended on one pound of water, the temperature of the latter had risen one degree.

The value of 772 foot-pounds is known as Joule's equivalent. More recent experiments have given higher figures, and the value of 1 Btu = 778.26 foot-pounds is used today.

Psychrometry

Psychrometry is that area of physics that involves the measurement and determination of atmospheric conditions, including the effects of moisture mixed with the air. In air-conditioning systems the dry air and water vapor composing the atmosphere are understood to be separate entities and each has its own characteristics. This water vapor is not dissolved in the air, but it serves to moisten the air.

Air and Water Vapor Mixtures

Water vapor is the gaseous form of water at a temperature below the boiling point of water. It is the most variable part of the atmosphere. At certain temperatures and barometric pressures it is extremely unstable in either gaseous or liquid form. This is shown by the rapid formation and disappearance of clouds and fog.

Water vapor makes up about 3% of the total air by volume in hot, humid weather and about 1/3 to 1% of total air by volume in dry, cold weather. Water vapor is steam at very low temperatures and it acts like steam at these low temperatures.

Dehumidifying

The air becomes humidified when moisture is added to it and is dehumidified when moisture is removed. The addition of moisture to air and the removal of moisture from air are important to air conditioning. An excess or deficiency of moisture has a noticeable effect on comfort and depends on the relative humidity.

Relative Humidity

Relative humidity is defined as the ratio of the quantity of vapor actually present in the air to the greatest amount possible at any given

temperature. The relative humidity of air at any given temperature can be found by dividing the amount of moisture actually in the air by the amount of moisture that the air can hold at that temperature and multiplying the result by 100.

Relative Humidity Measurement

Relative humidity has been traditionally measured by sling psychrometers. This instrument has two thermometers fastened to a frame. The instrument is spun around on a chain so it comes in contact with the maximum amount of air.

Around the bulb of one thermometer is a small piece of cloth, which is dampened with water before taking a reading. The evaporation of moisture from the bulb of the wet thermometer causes it to read lower that the one which is dry. The difference between the readings of the two thermometers provides the relative humidity.

Dew Point

The saturation temperature for a given quantity of water vapor in the atmosphere is called the dew point. At a given atmospheric pressure, the dew point is the temperature of saturation where the moisture begins to change into the form of tiny water droplets or dew.

Effective Temperature

Effective temperature provides an index of the degree of warmth or cold apparent to the human body. It takes into account the temperature, moisture content and motion of the surrounding air.

The effective temperature is not strictly a measure of heat in the same way that dry-bulb temperatures are. The effective temperature can be lowered by increasing the rate of airflow even though the wet- and dry-bulb temperatures remain constant. If the air movement is constant, the effective temperature is determined by the wet- and dry-bulb thermometer readings.

Dry-bulb and Wet-bulb Temperatures

Dry-bulb temperature is the temperature that an ordinary thermometer will measure. Wet-bulb temperature is the temperature at which the air becomes saturated when moisture is added without any change in the heat. If the bulb of an ordinary thermometer is surrounded with a moistened wick, placed in a current of air, and superheated with water vapor, the measurement obtained will be at some point below the dry-bulb temperature. This minimum reading is the wet-bulb temperature of the air.

Wet-bulb Depression

Outdoor air in the summer is rarely fully saturated, there is usually a respectable difference between the dry- and wet-bulb temperatures. This difference between the two temperatures is known as the wet-bulb depression. If the dry-bulb temperature reading is 85 and the corresponding wet-bulb temperature is 65, the wet-bulb depression is 20.

Temperature-humidity Index

This is an index of the human discomfort resulting from temperature and moisture. It is found by adding the dry- and wet-bulb temperature readings, multiplying the sum by 0.4, and adding 15. About 10% of the population are uncomfortable before the index passes 70, more than 50% are uncomfortable after it passes 75, and almost all are uncomfortable at 80 or greater.

Psychrometric Charts

The different methods of treating air can be shown on a psychrometric chart. The chart is a graphical representation of the fundamental mathematical relationship dealing with the thermodynamic properties of moist air entering and leaving a heating or cooling unit.

Three basic formulas are used to determine the following quantities:

1. The total heat to be removed from both the gases and water vapor.

2. The sensible heat removed from the air.

3. The latent heat removed in condensing the water vapor.

Total heat = cubic feet/minute × 4.5 (h_1 - h_1)
= Btu/hour

Sensible heat = cubic feet/minute × 1.08 db_1 - db_1)
= Btu/hour

Latent heat = cubic feet/minute × 68 (wb_1 - wb_2)
= Btu/hour

where, h_1 = total heat for entering air, Btu/pound
h_2 = total heat for leaving air, Btu/pound
db_1 = entering air dry-bulb temperature
db_2 = leaving air dry-bulb temperature
wb_1 = entering air wet-bulb temperature
wb_2 = leaving air wet-bulb temperature

Most heating and air conditioning calculations are made assuming sea-level pressure and standard temperatures, but high altitude or high temperature applications will require converting from standard conditions to those actually present.

Comfort Charts

Because of the possibility of shock to the human system, it is not desirable to have too great a difference between indoor and outdoor temperatures, particularly in the summer. Also, from the standpoint of physical comfort, low humidity with high temperature is comparable to high humidity at lower temperatures.

The greatest physical comfort in cold weather without undue high temperatures is achieved by increasing the humidity of the conditioned area. This also explains why some parts of the country are not extremely uncomfortable even though the outside air temperature is 100°F or more with a low relative humidity.

Heat Leakage

The thermal properties of a building affect the design of the air conditioning and heating systems. The rate of heat loss through walls, floors, and ceilings is important when calculating the heat and/or cooling load required for a building or area. Air leakage will take place through various apertures, such as cracks around windows and doors, through walls, ducts and chimneys.

Most of the components of the cooling load vary greatly during a 24-hour period and economic considerations should be the determining factor in the selection of equipment for heating and cooling season operations. Heat leakage is usually given in Btu per hour per degree Fahrenheit temperature difference per square foot of exposed surface (Btu/square foot/hour°F).

Estimates of the maximum probable heat loss for each room or space should include:

- losses through confining walls, floors, ceilings, glass, or other surfaces and
- infiltration losses due to leaks through cracks and crevices around doors and windows.

The heat leakage through walls, floors, and ceilings depends on the type and thickness of the insulating material used. The general formula for heat leakage is:

$$H = KA(t_1 - t_2)$$

where H = heat required

K = heat-transfer coefficient, Btu/square foot/hour/°F
A = area, square feet
$t_1 - t_2$ = temperature gradient through wall, °F

For example, suppose you wish to calculate the heat leakage through an 8-inch concrete wall having an area of 100 square feet if the inside temperature is 70°F and the outside temperature is 10°F.

The heat-transfer coefficient of a plain 8-inch concrete brick wall is approximately 0.50 and substitution of values in the above equation gives:

$$H = 0.50 \times 100(70 - 10) = 3000 \text{ Btu/hour}$$

Natural Ventilation

The natural forces that move air through and out of buildings include wind forces and temperature differences between the inside and the outside. Air movement can be caused by either of these forces acting alone or in combination, depending on atmospheric conditions, building design, and location.

The use of natural forces for producing ventilation can involve the following conditions:

- average wind velocity,
- prevailing wind direction,
- seasonal and daily variations in wind velocity and direction, and
- local wind interference by buildings and other obstructions.

Without obstructions to change the wind's direction, the movements of the airstream along with the pressure, remain constant.

Air leakage occurs through the building's cracks or openings. The cold and heavy air enters at low levels and pushes the warm and light air out at higher levels. The same draft takes place in a chimney.

Temperature-difference Forces

The draft in a stack or chimney is known as the flue effect. The flue effect of a stack is produced in a building when the outdoor temperature is lower than the stack temperature. It is due to the difference in weight of the warm column of air in the stack and the cooler air outside. The flow due to the flue effect is proportional to the square root of the draft head, and is given by the following approximate relation:

$$Q = 9.4A\sqrt{h(t - t_o)}$$

where

- Q = airflow, cubic feet/minute
- A = area of inlets or outlets (assumed equal), square feet
- h = height from inlets to outlets, feet
- t = average temperature of indoor air, °F
- t_o = temperature of outdoor air, °F
- 9.4 = constant of proportionality, which includes a value of 65% for the effectiveness of openings

The constant of proportionality can be reduced by 50% for a constant of 7.2 if the effectiveness of the openings is not as favorable.

Control Techniques

In air-conditioning systems, various automatic controls are used. Wet- and dry-bulb temperatures, which show the temperature and humidity present, are always used. A hygrostat is sometimes used on smaller installations.

Zone Control

Zone control involves a division of the air supply delivered into zones with varying requirements. A zone is an area among a number of areas, which are all conditioned by a single supply system. Each zone can be satisfied simultaneously with the use of either single- or multizone control, depending on the number and size of each zone.

Conventional single-zone units are installed one to each zone, which results in a simpler application. Single units are used when each zone is large enough to justify a separate unit and when the system can accommodate this number of units.

Multizone control is preferred when several relatively small zones can be grouped together and handled by one control. The purpose of the multizone arrangement is to divide the total air volume into a number of separate parts and condition each independently. A two-zone system is shown in Figure 1-1.

This can also be done with zoning dampers, which provide the necessary individualized control. The air outlet from the unit is parti-

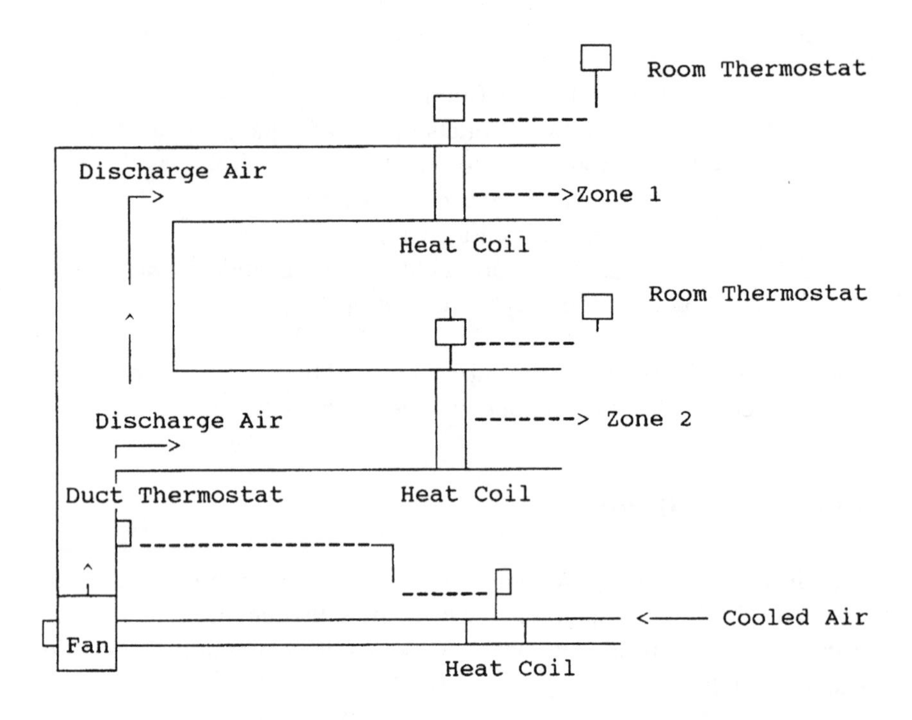

Figure 1-1. Two-zone, constant-volume, variable-temperature control.

tioned into a number of equal-size zones, each equipped with a set of dampers. The dampers regulate the desired final temperature in each of the zones.

Warm-air Heating Systems

A warm-air furnace is a self-contained and self-enclosed heating unit which may or may not be centrally located. Cool air enters the furnace and is heated as it comes in contact with the hot metal heating surfaces. As the air becomes warmer, it also becomes lighter, which causes it to rise. The warmer, lighter air will continue to rise until it is either discharged directly into a room or carried through a duct system to warm-air outlets located at some distance from the furnace.

After the warm air surrenders its heat, it becomes cooler and heavier. Its increased weight causes it to fall back to the furnace where it is reheated and the cycle is repeated.

A warm-air heating system is one in which the air is heated in a furnace and circulated through the rest of the structure either by gravity or motor-driven centrifugal fans. A system in which air circulation depends primarily on mechanical means for its motive force is called a forced warm-air heating system. Some gravity warm-air systems use fans to supplement the gravity flow.

Forced warm-air heating systems are often classified according to the type of duct arrangement used. The two basic types of duct arrangement used are:

- perimeter duct systems and
- extended plenum duct systems.

A perimeter duct system has the supply outlets located around the outer edge of the structure. Perimeter duct systems include perimeter-loop duct systems and radial-type perimeter duct systems. In the perimeter-loop systems, the ducts around the perimeter are connected in a loop while in the radial system there is no loop with only the radial feeder ducts from the furnace unit.

An external plenum system has a large rectangular duct which extends out from the furnace plenum in a straight line down the center of the building. Smaller supply ducts connect the plenum to the heat emitting units.

Year-around Air Conditioning

Year-around air conditioning systems warm the air during the winter and may also wash, filter, and humidify it. During the summer, the air-conditioning system will cool the air and perform humidification or dehumidification as required. A year-round air-conditioning system can eliminate the dust, soot, and germs from the air entering the intake ducts. It may increase the humidity in winter and decrease the relative humidity or moisture content of the air in summer. Circulation of the conditioned air also takes place. Figure 1-2 shows two arrangements of equipment for year-around air-conditioning.

Several types of service techniques can be used including gas-compression, absorption and steam-jet vacuum. Gas-compression is most common and consists of a refrigeration system employing a mechanical

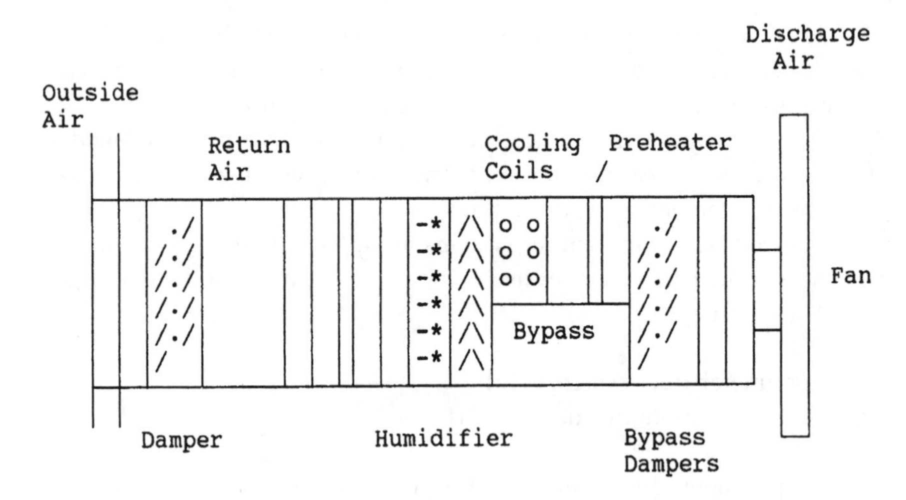

Figure 1-2(A). Central year-round air-conditioning system.

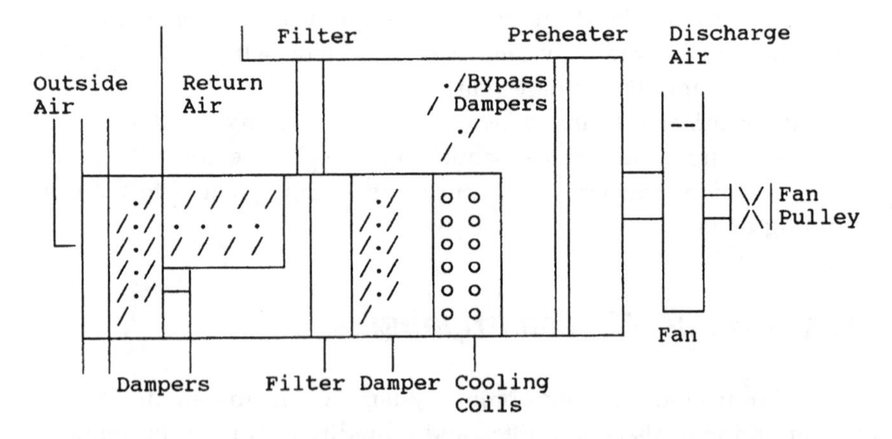

Figure 1-2(B). Central year-round air-conditioning system.

compression device (compressor) to move the low-pressure refrigerant on the low-pressure side of the unit and deliver it to the high-pressure side.

An absorption air-conditioning system uses a refrigerant gas in the evaporator which is taken up by an absorber and released in a generator upon the application of heat. A steam-jet air-conditioning system is a vacuum-refrigeration system where high-pressure steam is supplied through a nozzle and acts to eject water from the evaporator maintaining low pressure in the evaporator. The required pressure on the high side takes place by compression using a diffusion passage.

Central Systems

A central air-conditioning system is generally designed for assembly in the field rather than in the factory as a unit. One condensing unit can service several rooms at a lower cost than a number of self-contained units serving single rooms. Central air-conditioning systems provide air ducts to the various rooms along with one or more exhaust fans to remove any undesired proportion of the air.

Mixing

A mixing chamber is usually incorporated in the system for the purpose of recirculating a portion of the air within the building in order to reduce the load and operating costs. In the summer the cooled air and in the winter the heated air is mixed with the fresh air drawn in from outside, so that the load is reduced considerably.

Some fresh air is drawn in and mixed with the recirculated air. This is accomplished with a mixing damper. The mixing chamber must be airtight so that little or no air leakage occurs.

Heating Coils

In cold weather regions a heating or tempering coil is used to warm the intake air. This is used when the outside air is at such a low temperature that the heat from the recirculated air may not be sufficient to meet the requirements of the load.

Tempering heaters must be used wherever cold conditions are found. In some cases, these conditions may not occur every year or for any length of time. The tempering heater is designed to prevent cold incoming air from causing the water spray to freeze.

There are several different types of heaters available. The electric heater requires little space but is expensive to operate unless electricity is available at very low rates.

Preheaters

A preheater may be used to add heat to the airstream when the temperature of the incoming air is so low that the tempering heaters on which the heating load falls are unable to maintain the required 68 or 70°F at the fan.

The preheater is used only where extreme temperature conditions are encountered. In many systems, the tempering heaters are made with enough capacity to carry the heating load.

Cooling Coils

The cooling coils may be chilled by means of refrigerants or a pump-forced liquid, such as brine or water. If water of the proper coolness is obtainable in sufficient quantity, as is the case in many northern cities, it may be employed, since it requires only a pumping operation to lift it out of the ground. Cold water can also be obtained with mechanical refrigeration equipment.

Air-supply Fans

Supply fans and blowers are available in various types for specific purposes. The selection of the proper type is important if the operation is to be economical and efficient. In general, fans can be divided into two types: propeller and centrifugal or turbine.

Propeller fans are used in short ducts or when there are no ducts. The low speed of the blades near the hub, compared with the peripheral speed, makes this form of fan unsuitable to overcome the resistance encountered in long ducts. The air driven by the effective blade areas near the rim can pass back through the less effective blade areas at the hub than to move against the duct resistance. Propeller fans also tend to create more noise.

Many air-conditioning installations require long air ducts to carry the conditioned air to the various areas of the building. Centrifugal and turbine-type fans are employed in these installations.

Filtering

Air filters include those that can be mechanically cleaned as well as replaceable and electronic units. The air filter is usually operated year round. There are times when the cooling and humidifying functions are not needed, but the filter is one part of the system that may be kept in continuous operation.

Air Washers

An air washer is a unit that brings the air into contact with water or a wet surface. The principal functions of the air washer are to cool the air passed through the spray chamber and control the humidity.

In some cases, the air may be passed over the wet surface. More intimate contact is possible when the water is divided into a fine spray and the air passed through the spray chamber.

Some of the finer water particles will be carried along with the air current and a series of curved plates called eliminators may be used to change the direction of air flow and throw off some of the water particles.

The water is generally circulated with a pump. Warm water is passed over refrigerating coils to cool it before being passed to the spray chamber.

In some cases, the water is cool enough to use as it is drawn from the ground. In other cases, the water is not cool enough and must be cooled by refrigeration.

Indoor Design Conditions

The indoor conditions to be maintained in a building are the dry-bulb temperature and relative humidity of the air at a breathing line 3 to 5 feet above the floor. The area should indicate average conditions and not be affected by abnormal or unusual heat gains or losses from the interior or exterior.

Since it is difficult to maintain the mean radiant temperature at the dry-bulb temperature in actual practice, the effect of warmer walls normally encountered during the cooling season requires a lower design condition. A 75°F design dry-bulb temperature is often used which is slightly lower than the optimum range of 76.5 to 77.6°F.

During the heating season, the mean radiant temperature is normally lower than air temperature, except in some panel-heating installations where the mean radiant temperature can be higher than air temperature. This suggests a higher dry-bulb temperature than 75°F for comfort.

When unusual application conditions exists, such as large areas of glass and inside partitions with widely different temperatures, an adjustment of the design dry-bulb temperature may be necessary to maintain comfort conditions. When the mean radiant temperature condition of 75°F is applicable, the dry-bulb temperature can vary from 73 to 77°F in the occupied zone.

The air temperature at the ceiling can vary beyond the comfort range. A nominal 0.75°F increase in air temperature per foot elevation above the breathing level can be expected in normal applications, with an approximately 75°F difference between indoor and outdoor temperatures.

For summer cooling, a relative humidity of 50% is average for the United States. In arid climates, a design relative humidity of 40% or less is more realistic. During winter conditions, the relative humidity maintained in the building depends on the severity of the outdoor conditions. The relative humidity should not exceed 60% in the occupied space to prevent material deterioration and normally should not go below 20% to prevent human nostrils from becoming dry and furniture and other items from becoming too dry.

Weather Station Interpolation

Data from weather stations at specific locations and at published elevations furnish a system from which, by interpolation, estimates can be made of the expected conditions at a precise location. For a lower elevation, the design values should be increased, while for a higher elevation, the values should be decreased. The increments normally used in these adjustments are:

Dry-bulb temperature: 1°F/200 feet
Wet-bulb temperature: 1°F/500 feet.

Adjustment for air-mass modification may be needed near large bodies of water where air moves from the water over the land in the summer. On the West Coast, both dry- and wet-bulb temperatures increase with distance from the ocean. In the region north of the Gulf of Mexico, dry-bulb temperatures increase for the first 200 or 300 miles, with a slight decrease in wet-bulb temperature due to mixing and drier inland air. Beyond the 200- to 300-mile belt, both dry- and wet-bulb values tend to decrease at a somewhat regular rate.

The difference between large areas of dry surfaces and large areas of dense foliage upwind from the site can account for variations of up to 2°F wet-bulb and 5°F dry-bulb temperature. The warmer temperatures are associated with dry surfaces.

Weather-oriented Design Factors

The general approach to air-conditioning system design involves the computation of a peak design load at a condition established using one of the frequency of occurrence levels of dry- and wet-bulb weather data.

During the four warmest summer months, these values occur between 2:00 p.m. and 4:00 p.m. The winter statistics occur during the 3 coldest months between 6:00 a.m. and 8:00 a.m. The daily dry-bulb variation will be of the order of the daily range.

Statistically, the daily range is the long-term average daily range for the warmest month. The daily range is generally greater during clear weather and much less during cloudy or rainy weather.

An approximation of the distribution of the hourly dry-bulb temperatures can be made by using the daily maximum at 3:00 p.m. and the minimum at 7:00 a.m.

In the continental or inland areas of the United States, the maximum dry-bulb and maximum wet-bulb temperatures are not coincident. In maritime areas they tend to be coincident.

The assumption of dry- and wet-bulb coincidence can result in weather-oriented loads of up to 1/2 greater than might be expected. One solution is to determine if the structure is most sensitive to dry-bulb (extensive exterior) or wet-bulb, (outside air ventilation).

From several years of local weather records, select days with the maximum wet- and dry-bulb entries. Tabulate the dry-bulb peak with the corresponding wet-bulb and the wet-bulb peak with its corresponding dry-bulb. This will show the local coincident data.

As more complicated and sophisticated applications are utilized to meet energy-saving demands, design maximums based on peak values will not suffice. Off-peak values need to be considered. Some types of days that must be accommodated are more frequent in occurrence than maximum or minimum design days. Examples of these include cloudy with small temperature change, warm a.m., cool p.m. and windy.

Machine computations allow rapid calculations of dry-bulb temperature with its corresponding wet-bulb maximum for several hours daily on both a room-by-room and a zone or building basis. This can result in more accurate control as well as more compatible system operation. It can also provide energy-consumption estimates.

Humidity Controls

Humidity control involves controlling the amount of water vapor present in the air in the space. When the relative humidity at the desired temperature setpoint is too high, dehumidification is used to reduce the amount of water vapor in the air. When the relative humidity at the desired temperature setpoint is too low, humidification is used to increase the amount of water vapor in the air.

The relative humidity depends on the dry-bulb temperature. Dry bulb temperature and relative humidity are usually specified concurrently. A 70°F dry bulb and 50% RH measured in a room has an actual moisture content or specific humidity of 54.5 grains of water per pound of dry air. A room with the same specific humidity at 60°F will have about 70% RH. At 80°F this becomes about 35% RH.

Humidity controllers are also known as humidistats. These units are usually contiguous to the thermostat so the humidity is based on the ambient temperature. The humidity controller is set at the relative humidity setpoint. A change in relative humidity from the setpoint causes a control signal to be sent to the controlled component.

Humidifiers use water spray, steam grid and steam pan techniques. In a water spray system, the controlled component is a water valve.

If a duct-mounted steam grid humidifier is used, the control signal opens the steam valve at the inlet to the humidifier unit. As the steam valve opens, steam flows through the humidifier in the supply airstream and raises the relative humidity.

Dehumidification methods involve the use of desiccants, surface cooling and indirect cooling coils. Indirect cooling coils are used with sprayed coil dehumidification.

Dehumidification may take place with sensible (dry bulb) cooling using surface dehumidification on the cooling coils. Dehumidification in low dew point process systems is usually done in a separate dehumidification unit.

Air leaving the cooling coil from surface dehumidification is usually near a saturated condition. If cooling is controlled from the relative humidity in order to remove water vapor, the supply air will often be cooled more than is required for sensible or dry cooling. The space may then require some reheating to prevent overcooling.

HVAC Intelligence

Most improvements to HVAC systems involve increasing their intelligence. Setback thermostats, which are also known as clock or electronic thermostats, automatically raise or lower indoor temperatures to meet the building's schedule. Indoor temperatures can setback to lower than average comfort levels while the building is unoccupied, then they are automatically brought back up to the comfort level before it is fully occupied. Other smaller setbacks might start at mid-morning when the outside temperature starts to warm.

Without any sacrifices in comfort, a 10 to 20% savings on annual heating bills are feasible. In the summertime the setback feature is used to save on air conditioning costs. Some clock thermostats use 7-day programs with a different schedule for each day of the week. Another schedule is the 5 + 1 + 1 weekly program, which uses a Monday-through-Friday timetable and separate timetables for Saturday and Sunday. There are also 5 + 2 schedules which use two programs, one for the five workdays and another for the weekend.

Programmable units have evolved into flexible electronic units with displays and buttons from the earlier electromechanical models that used small levers or pins to mark the setback times. Manual overrides are used to override the current temperature setpoint without reprogramming the thermostat. Newer models automatically return to the original program after the override period is complete. Electronic thermostats also use a key or switch that allows you to change from a heating season program to cooling, or back again without reprogramming. Some models will make the changeover automatically.

Heating-only and cooling-only systems have two or three wires, and combined heating-cooling systems typically have four or five wires. Heat pumps use eight or nine wires.

Electronic thermostats use batteries to maintain the program in case of a power failure. Most units use a 24-volt power source, but there are also 120-volt or millivolt systems. Some use batteries as their only source of power. This type of battery-only thermostat should have a low-battery indicator to warn of impending problems.

A key improvement in the future is to equip systems with more and better sensors so that the controller has the proper feedback it needs to make smarter adjustments. These sensors will monitor not only the indoor and outdoor air temperature, but relative humidity and wind conditions may be monitored as well.

More extensive climate control will be used that defines heating and cooling needs using current weather information along with forecast information in some cases.

The concept of zoning will also grow as a cost effective technique and control systems will use more thermostats and/or remote sensors throughout the building.

Pressure Control

If a forced-air system is not in pressure balance, some areas will be pressurized while others are depressurized. The depressured areas will be subject to intrusion by cool, dry air from outside. Ductwork leaks allow the depressurization effects and dryness to become worse. When the ductwork is used for air conditioning, depressurization can allow hot, humid air to enter. When a furnace draws its combustion air from outdoors, this can also cause depressurization.

One way to reduce the dryness and fuel consumption is to raise the relative humidity inside. Another technique is to add new supply or return air ducts to supply a better balance or regulate the system by opening and closing air outlets. Air outlet and duct control in the future will be based on a continuation of existing trends includes increased integration, better communications, more intelligence in the equipment, increased use of PCs and improved price/performance ratios.

Motion Control Integration

Integration of motion control systems and the information that they control in future automated control systems will make this technology more useful, less complex, more powerful, and physically smaller. As the machine adapts to use fewer and smaller components, new forms of actuators will be adopted.

Smarter and smaller electronics will allow multi-box solutions to become single-box solutions with the electronics built into the motor. With the trend towards more open controls, the drive intelligence will be integrated further into the motion control module.

The first step is the integration of the control with the amplifier. Intelligent drives will replace power blocks and traditional amplifiers. An intelligent drive contains the traditional amplifier along with features like auto tuning, auto tuning on the fly, diagnostics, horsepower calculation, calculation of the three main torques being delivered (friction torque, acceleration torque, and work torque), feedback of position, velocity, acceleration and inertia calculation. Once you know acceleration torque and acceleration rate, inertia can be calculated.

There are already products that can do closed-loop control with a hybrid step motor. In the future, motion control platforms will allow any motor type to be controlled for a wide range of velocity and positioning applications. The feedback device may be an encoder, resolver or tachometer which will be electronically recognized and controlled to provide the needed feedback signals. Many motion applications will use sensorless positioning techniques where embedded signals from the motor itself provide the feedback signals.

Contrex has a line of position control products designed to add intelligent control and operator interface to all types of variable speed drives. This makes it possible to register and synchronize independent machines, allowing them to work together and to follow preprogrammed motion profiles. Rockwell Automation has added enhanced motion control in their Allen-Bradley ProcessLogix distributed process control system.

Feedback Devices

While more sensorless systems may be appearing, many motion control systems will continue to need feedback devices. These are mostly encoders and there will be continuing refinements in these. Incremental encoders will evolve into networkable smart devices that take over tasks now done by absolute encoders.

Since they will be tied into the network, they will always be powered. If a position is changed when the machine is powered off, they will still be able to monitor position and the machine control server will be able to maintain an absolute position.

The newer encoders can also function in temperatures that traditionally have called for resolvers. BEI Motion Systems, for example, has encoders that operate at 105°C, and Admotec has magnetic encoders that work from -30 to 90°C.

Sometimes a feedback device is not used. Semipower Systems' IndexBlok drive module can position 1 to 20 kW three-phase motors with up to 16-bit resolution per electrical cycle of the motor with no feedback device. It gets its information from the motor itself and processes it with special software.

Linear encoders continue to improve. Danaher Controls' Dynapar LR/LS inductive linear encoder provides resolutions to 0.25 micron without the use of glass scales.

Feedback devices will undergo more changes besides the addition of network capability. Conventional encoders will be replaced by laser sensors and fiber-optic sensors in the future.

Trends in Motor Technology

The move to integration involves building all the electronics into the motor. Animatics has done this with its SmartMotor line. The integration of controls into the motor package will be paced by advances in power control devices. Smaller, more intelligent and more efficient motor controller packages will operate both brushless and stepper motors.

Motors and the mechanical components associated with them will gain in performance and decrease in price. This will be due partly to improvements in materials, especially magnetic materials and insulation systems. Servo technology is becoming more easier to use and tune.

Brushless servo motors and amplifiers will show a continuing decline in price, size, and increases in performance. This will cause them to dominate the market at the expense of stepper motors. Stepper motors will still have their applications due to stepper drive technology innovations.

Stepper motors are improving and they have their advantages and disadvantages. They initially became popular because they ran open loop and were simple to apply. No feedback loop was needed. The controller just sends the proper number of pulses and then goes to this position. Once in position they do not hunt, as servos can. However, they have some peculiar dynamics, and under certain conditions and speeds can become unstable. If they miss steps, the system is out of position until reset.

Microstepping is a way to increase the number of positions that a step motor can assume. It has been refined by the Compumotor Division of Parker Hannifin in its OEMZ14 microstepping drive. This drive uses special damping circuits for Anti-Resonance and Electronic Viscosity damping. These are designed to defeat the step motor's resonance effects and stop any ringing that can occur after quick moves.

There are several approaches to the missed-step problem. Oriental Motors' Alpha Step system uses an integral encoder and automatically switches to closed-loop control if the motor begins to lose synchronization. When the lost steps are made up, the system switches back to open loop.

Compumotor uses another approach to the synchronism problem. Its Gemini series can recognize if the motor has stalled without a feedback device, and the method is being extended to allow the step motor to operate closed loop.

Linear motors are gaining in performance and popularity, and this trend is likely to continue. Their ability to eliminate lead screws, gears, pulleys, and the associated complexity and backlash will increase their use.

Linear motors are increasing in power, precision, and ease of use, while prices are decreasing. This will continue as linear motors prevail as the fastest growing motor segment.

Networking

Integrating motion control involves more than moving the control and drive into the same package or even moving them into the motor. Integration starts with communication and is focused on open systems.

Motion control equipment has used the SERCOS standard for some time. There is an increasing movement to DeviceNet, Profibus, MODBUS, and others including USB and IEEE 1394 (FireWire). DeviceNet is used by Whedco which is a subsidiary of GE Fanus. The trend to distributed control uses open architectures such as DeviceNet.

DeviceNet is popular but it has some limitations. Since it is a device-level bus, it was not designed for transferring larger files. File transfer capability is being added with a new DeviceNet System file transfer specification. This should allow the motion controller to be run from the DeviceNet ports without the need for an additional serial port.

The use of Ethernet for motion control applications is spreading and should continue. It offers technical and ease-of-use advantages. Ethernet will be used in the future as a bus for motion, I/O, and sensors. Some work still needs to be done at the application layer, to allow interoperability. Ormec Systems and machine vision supplier DVT have created a vision-guided motion control system that uses Ethernet TCP/IP to connect the vision system and motion controller. Ethernet will continue to grow in importance, but Profibus-DP will also gain in acceptance for motion control.

Universal Serial Bus (USB) is finding industrial uses, especially for data acquisition. USB for motion control is also appearing with a USB-connected drive by Galil Motion Control.

Another bus that is moving in the industrial world is the IEEE 1394 FireWire. It runs at 200-Mbps, which is fast enough to be inside a servo loop. It is plug-and-play and uses low-cost cable and connectors. Ormec Systems has a version of 1394 for servo applications called ServoWire. Acroloop Motion Control Systems is also using IEEE 1394.

In applications that require axis coordination for path control, SERCOS has established itself as the dominant open network with the intelligence distributed to the servo drives. In simpler applications where axes movements are independently sequenced or a number of axes are electronically geared together, more intelligence and control are distributed from the servo controller to the drive. These smart drives will then be connected together via a field bus.

PCs

PCs have moved into more industrial roles. Starting as programming devices and operator interfaces (OIs) for dedicated control equipment. They have taken over more control tasks from DCS to soft PLCs.

PC use has steadily expanded in spite of persistent worries about nondeterministic behavior.

Just a few years ago, PCs were used as motion programmers and as OIs, while the actual motion control was done by a dedicated unit that drove a separate amplifier. Then, motion control cards were plugged into the PC's backplane. The next step is to let the PC do the motion controller task in software, using its own real-time operating system.

PCs will take over an increasing part of the data handling and communications function, helping to tie the motion control function into the entire system or network. Dedicated, proprietary boxes will be in the past as open controls become the future. As PC-based controls dominate, there will be a continued trend towards more general-purpose controllers. As these general-purpose motion controls start to dominate, costs will be driven downward through better manufacturing techniques, including reducing the controller to the chip level.

PC-based control software has achieved development time savings due to a single programming environment, single database, and full integration with logic and HMI (human-machine interface) functions.

Continued advancements in mainstream PC technology include faster processors, better operating systems, advanced client/server computing architecture, new communications technologies and computer-based tools for test, measurement and industrial automation.

The PC allows measurements to be easily sequenced, processed, displayed, and stored. Seamless integration of motion is fundamental to the success of automated systems. The PC is the viable platform for these integrated systems.

There are PC-based control system that include PLC functions to fieldbus I/O, motion control via SERCOS, and communications via serial ports and Ethernet. In the future this could all be done more simply with plug-in Java Beans for motion control or for the interface between motion control and the HVAC system.

Motion control will become a software function running on a real-time operating system in a PC, at first under Windows NT, and later perhaps under CE. The control program will be a combination of soft-motion and soft-logic technology. Logic programming will be standardized by IEC-1131-3 while motion programming may continue to be vendor specific until a similar standard is developed for motion controllers. The PC will then be enabled to take a greater role in motion control.

Integrated motion controllers based on Windows NT are starting to appear. One system uses an industrial Pentium-class PC running NT and Steeplechase Software's Visual Logic Controller.

As faster processors are developed, and Microsoft continues to add features to their operating systems, you will see the advent of soft motion control performed in CE and NT platforms. The increased use of PCs will create easier data movement and field bus architectures will remove any separation and create truly integrated systems.

Advanced software technologies such as ActiveX and graphical programming software such as LabVIEW, will continue to make motion control applications more intuitive and simple.

Besides better damper positioning, tightening the exterior envelope will slow the passage of cold, dry air from outdoors to in and raise the relative humidity level inside. You tend to feel warmer when the relative humidity goes up, even if the air temperature has not. Providing the furnace with more outdoor combustion air will stop the furnace from depressurizing the rest of the building and reduces any backdraft. This can result in energy savings, since the furnace is no longer using previously heated air for combustion.

Forced-air heating has its own comfort problems. The air may be forced out at 90°F or warmer over people's skin. This has a cooling and drying effect. The moving air results in convective heat losses [cooling] from the surface beneath it as well as evaporation off the skin. One way to reduce this discomfort from moving air is better air control with dampers and fans. Some oil and gas furnaces have reduced outlet air temperatures and increased air flow for efficiency reasons.

Zones

Zoning involves heat delivery only to those areas where it is needed and when it is needed, so waste is kept to a minimum. Many early zone forced-air systems were hard to balance and control. The advent of low cost microprocessor controls and motorized dampers to modulate airflow has allowed forced-air zoning to be practical.

Trends in Damper Control

In the past, proprietary motion-control hardware has made interoperability among system components almost impossible. This is changing as standards like SERCOS are making it easier to link computers and damper controls. This linkage is occurring because of the growth of C/C++ development tools, standard Windows microcomputers and the serial real-time communications (SERCOS) standard. SERCOS was first adopted in Europe and its use started in the United States in 1995. SERCOS is designed to establish communications between digital-motor drives and controllers.

SERCOS defines a basic way to plug in up to 254 programmable drives into a 4-Mbps fiber-optic ring that functions as the medium for drives and control to pass information serially to each other. SERCOS provides the two-way communications between programmable drives and controls.

The SERCOS fiber-optic ring becomes the medium for controls and drives to pass data serially to each other. The type of fiber specified for SERCOS is a nonstandard diameter that must have a screw-on connector rather than the more conventional bayonet clasp.

Unlike analog interfaces, SERCOS allows two-way communications between controls and programmable drives. It represents a plug-and-play bus for these control systems. SERCOS eliminates bugs that were common in the wiring between the controller and programmable motors.

The open architecture for motion-control systems results from teaming the SERCOS standard with PCs and standard development tools. PCs are more economical than programmable logic controllers (PLCs) which is a likely alternative. Standard development tools diminish the frustrations of using proprietary systems.

Digital Drives

Motion-control is moving from analog to digital components, particularly digital programmable drives. Digital drives are faster and more precise than analog. They can go from 0 to 3000 rpm in 20 milliseconds or come to a stop within 1/8000 of a revolution.

SERCOS is not the only interface that can be used for motion-control systems. Besides proprietary systems, the alternatives include Fieldbus, Profibus, CANbus, InterBus-S, and DeviceNet. These establish a communications link for I/O data to travel to a number of field devices. DeviceNet has been used for I/O control and SERCOS for motion control in the same system.

SERCOS can lessen wiring costs. An analog-motor drive can have 16 wires. Simplified wiring means less wiring errors. Networking drives and controls through the SERCOS interface is a straightforward matter. With the ability to manipulate every variable in a drive system, SERCOS digital drives have brought drive setup and optimization into the graphical environment of Windows.

Although the standard allows a maximum of 254 programmable-drive connections in each fiber-optic ring, the actual drive limit can be less depending on the communications cycle time and how much data must be passed for the application. The cycle time is also a factor if these systems are used in building applications.

SERCOS was designed for synchronous applications. The actual values are detected and commanded values acted on by all drives simultaneously. The drives work synchronously, so beat frequencies or harmonics are eliminated.

Suppliers of SERCOS motion-control cards and equipment include Indramat, Motion Engineering, and Pacific Scientific. New SERCOS systems may be capable of 10-Mbps rates. Digital technology means motion-control systems can be more precise and easier to use and SERCOS represents a future away from proprietary systems.

References

Anderson, Edwin P., *Air Conditioning: Home and Commercial,* Macmillion Publishing Company: New York, New York, 1985.

Brumbaugh, James E., *Heating, Ventilating and Air Conditioning Library,* Vol. 1, Theodore Audel and Company: Indianapolis, Indiana, 1986.

Chapter 2

Energy and Power Management, Distributed Control Trends

The late 1980s and the early 1990s were not the best of times for starting advanced distributed control projects. This period was characterized by slow or no growth and recession. Industry responded by cutting costs and drastic reorganizations. These took the form of mergers with plant closings or cutbacks, layoffs and delayed purchases for capital equipment.

The reduction of personnel put some pressure on the surviving departments to increase automation and become more efficient. The financial staff analyzed operations more closely and offered areas that might be improved. These economic factors as well as technological advances in electronics and control hardware allowed changes that were not possible before. Among the benefits of this new technology were tools that allow users to document procedures and justify them for the next budget year.

The mission to reduce costs did not mean quality could be sacrificed. Quality needed to be improved since quality expectations and standards existed because of regulation, market competition and litigation.

Deregulation Issues

Power deregulation is a reality in several states such as California and pilot programs are going in several others. The way energy is bought is changing rapidly. Energy deregulation offers great potential for cost savings.

Utility deregulation is a direct result of the Federal Policy Act of 1992. Competitive market-based pricing is replacing state and federal rate structures. In states that are still regulated, utilities are modifying their rate structures to preserve their customer base in any future deregulated environment.

Open, competitive energy markets are unrestricted by geographical boundaries and regulated rates. The different purchasing options and rate structures are similar to what occurred following the deregulation of the telephone industry.

Advances in metering hardware, communications, and software have significantly reduced the cost of how to monitor and control energy use, even in regulated environments. Those who delay using these advances may see their costs rise significantly.

Further advances in software and communications allow facilities to link their energy costs with labor/material costs and production rates. These new tools and technologies will allow companies to negotiate better rates with utility suppliers and determine more efficient production methods and schedules. These companies will be able to find many ways to significantly lower the once fixed cost of their energy use. The new options also allow companies to guard against unexpected power reliability.

Understanding energy requirements requires that you must know how your operations can tolerate an occasional interruption of power. Production or other operations may need to be shifted to off-peak times. Partial load interruptions or lower power quality may damage some equipment and expand maintenance costs resulting in lower productivity and lower quality.

In recent years, environmental and political pressures have forced more requirements on companies that discharge waste of any type. Many industries are also facing higher quality standards. One example of higher quality are those standards enforced by the U.S. Food and Drug Administration (FDA) for pharmaceutical manufacturers.

High levels of competition are also forcing improved quality. Competitive factors and improved technology mean previously acceptable levels of product quality become unacceptable. Most industrial customers require suppliers to have International Standards Organization (ISO) certification in all manufacturing processes.

Legal Ramifications

The legal area continues to grow. Many activities that were conducted easily without problems now require careful justification and documentation. The trend is towards more detailed recordkeeping to help a company prove its case if needed at a later date. Building owners as well as manufacturers have seen an increase in litigation, in areas such as indoor air quality.

Improving quality and holding down costs is proving to be a difficult challenge for those in charge of operating a building. Building efficiency requires capital investment. When money is limited, operations and maintenance functions can fall out of the bottom of capital funding requests. These cost-avoidance projects are often passed up for more lucrative, but possibly riskier, direct revenue-producing projects. Since departments must compete among themselves to receive the money they need, better documentation of operations is needed.

Energy Management and Networks

Networked control systems started during the energy crises of the 1970s, when the rising prices of imported oil triggered severe restrictions on energy use and prompted more efficient energy management and control. This resulted in the development of energy management systems (EMS) for tightly monitoring energy usage. These systems grew over the years in both sophistication and scope.

One offshoot appeared in the 1980s called building automation systems (BAS). These systems added historical data, trend logging and fire and security functions to traditional energy management functions. These applications focus a return on investment based on utility savings.

Analog and Digital Systems

The advent of direct digital control systems in the mid-1980s displaced older analog closed-loop control (Figure 2-1) as the temperature control scheme of choice for most large equipment. The digital systems improved both accuracy and reliability. But, these systems were modeled after existing system architectures that did not include intelligent, stand-alone field devices. There were still numerous interfaces to the various building systems and the major decisions were made at a central computer.

Integrated Systems

More recent Building Automation Systems (BAS) attempt to limit the interfaces and provide a more seamless, integrated network. Digital control networks provide an architecture that can be fully distributed with independent controllers for the systems and subsystems in a building. Ideally, all of the various components will talk to each other in a common language.

The networked system allows improvements in plug-in instrumentation, instant decision support, documentation and automation. Intelligent, versatile instrumentation at the lowest control level allows the energy management data to be easily organized for flexible efficient management.

Highly organized data collection means the facilities and operations staff can be more effective. The reports allow better decisions since

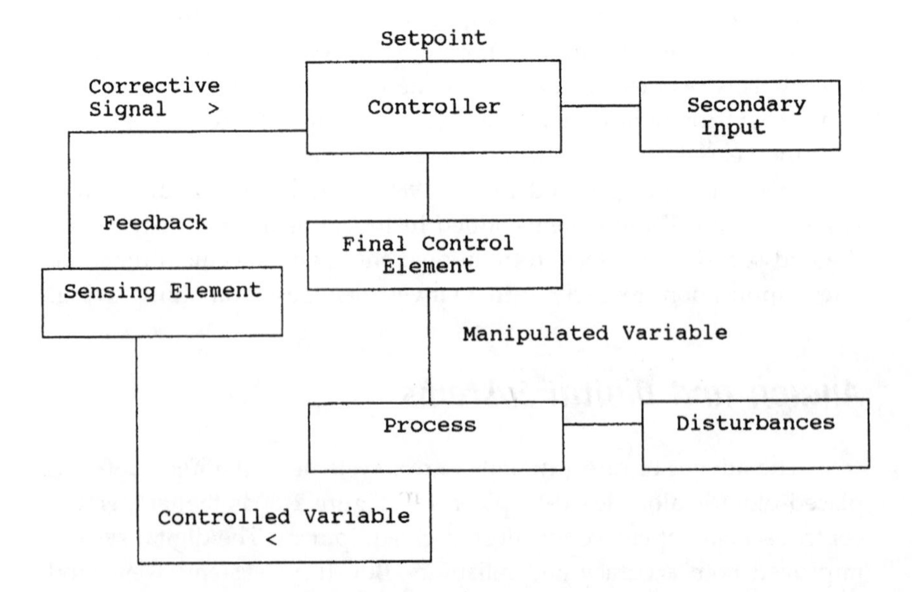

Figure 2-1. Analog control closed-loop system.

a variety of reports can be obtained quickly. Audit trails also go more quickly and can be used to substantiate proof of performance, as well as limit future risk. A networked control system can automatically document processes, generate reports and issue work orders when problems are detected at a point in the network.

Levels of Control

Networked control systems generally use several levels of hierarchy in their distributed architecture. Three levels that can be used are:

- distributed controllers,
- building-wide control and
- information management.

Each level serves its own purpose, but all levels are interconnected, similar to the operating structure of a corporation.

Distributed controllers may employ microprocessor-based sensors and other devices to meet the needs of a specific application. These stand-alone controllers or control systems function as specialized tools for a specific job.

The building-wide level coordinates all of the building control strategies. This level coordinates the specialized activities and provides global direction.

At the information management level, data collected from various points in the system are transformed into usable information. The combined efforts of the highly specialized field units the higher levels of control to provide and in some cases make information decisions about the operations of the overall system.

Distributed HVAC Control

Distributed control refers to a complete control system with all the needed inputs, output and control processing logic to affect a control loop. In HVAC the controlled parameters include standard functions such as discharge air temperature, space temperature, humidity and fan control.

The benefits of such a control system in an intelligent, integrated network include repeatable and individual parameter or area (zone) control. Specific comfort zones need consistent and repeatable results as loads switch from occupied to unoccupied. The system must switch back to the same comfort conditions for the next occupied mode. It is possible to control a specific room or area at the appropriate level. Intelligent devices can be tied into the network, allowing data to be collected and energy usage to be measured.

Individual comfort control can increase employee output and provide annual productivity gain of over $1000 per employee. A networked system also offers significant cost savings compared to a hard-wired system as shown in Table 2-1.

Table 2-1. Hard-wired Versus Networked Systems

Hard-wired System	*(424 I/O Points)*	
Labor and Material	385,000	908 I/O Point
Engineering/Design	75,000	176 I/O Point
Total Cost	$460,000	$1084 I/O Point
Networked System	*(424 I/O Points)*	
Labor and Material	200,000	471 I/O Point
Engineering/Design	35,000	82 I/O Point
Total Cost	$235,000	$ 553 I/O Point

Networking takes building automation beyond traditional HVAC functions. A networked system can also manage lighting, fire and access control. If these systems are fully integrated with HVAC, then the expanded integrated control functions can also address environmental issues such as indoor air quality sequence or CFC leak alarming.

Communications

Higher levels of integrated control are possible by tying together distributed controllers in a communications network. This allows a reporting path that allows information to flow from one controller to another.

Using coordinated control sequences, the entire building automation system can be monitored and its various functions optimized. All of this can take place transparently, behind the scenes, automatically.

Information Management

This is the highest level of control in the networked system. Data from hundreds or thousands of I/O points in a building or building complex can be accessed quickly and used to assist in decision making.

The proper communications architecture is needed so that easy access to system information can take place at these different locations throughout the facility. The access could take place at a local or remote personal computer workstation or terminal.

Information management is needed to provide both regulatory compliance and risk management. Financial decision making is also allowed along with quality assurance. Automation allows a speedy response to problems, as well as their resolution. A maintenance management feature can also issue and track work orders. Networked control provides quality assurance which can be used to identify, analyze and improve building operations related to both comfort and security.

A large part of the building's set of plans can be loaded into the computer. For new buildings this involves access to the CAD (Computer Aided Design) system that designed the building. For older buildings the drawings can be scanned into the CAD system and then utilized by the Energy Management System.

Documentation is often needed for regulatory compliance. This documentation may include testing, proof of performance, and incidence reporting. It is essential in managing and reducing risk. Historical data can be used to identify cost-saving opportunities.

Hard-wired and Networked Systems

The earliest Energy Management Systems used devices that were hard-wired back to the computer. A distributed format evolved that used multiplexed signals over a common wire or the electrical distribution system (power line carrier systems). These power line carrier systems suffered some setbacks from early reliability problems.

Multiplexing reduced the cost of wiring from remote panels to the computer but did not reduce the cost of wiring the input/output devices. Multiplexed systems have a reduced response time as the system gets larger. This response time becomes less important as distributed systems put more computing power out in the remote panels and even in the input/output devices themselves. The speed of information transfer can be increased by switching from twisted pair cables to coaxial or fiber-optics, however, these types of cables add to the installation costs.

In the future, communications between sensors and multiplex boxes and the rest of the system may use a combination of technologies including traditional means such as twisted wire and coaxial and non-traditional methods such as infrared or radio wave.

Operator Interfaces

The first control panels used individual pilot lights, then came single line light emitting diode displays. The next evolution in control interfaces came with text only, monochrome CRTs. Today, high resolution color graphics provides users with realistic images that are updated once a second.

Multimedia computers can combine video, CD-ROM, audio, text, pictures and sound to enhance communications between the computer and operator. The system may customize the interface for each operator, anticipating what the operator wants to know.

A combined display/sound interface may be used which may mimic what we call virtual reality. In virtual reality, you can experience the environment. Special headsets, gloves and even full body suits are used.

The operator can experience the building in operation. After a complaint of a hot or cold temperature or a draft, an operator can zoom in to the space to feel and measure the temperature.

Zooming inside the VAV box, the operator could check the damper position and view readouts of air volume and temperature. The thermostat or damper control could be adjusted while observing the system's operation. The operator could also check the operation of fans, boilers and chillers using this zoom control.

Adding a sensor to a room could be a simple operation. The sensor might have a self-adhesive backing and stick to the wall, where power is supplied to the unit by a built-in solar cell with battery backup. The sensor would then start broadcasting using infrared, radio wave, or microwave. The computer will recognize the sensor and assign a point number.

The system will map the location of the sensor using triangulation of the signal and its internal map of the building. The system can then start a self-optimization routine to search for the optimum control strategy to utilize this new sensor.

Direct Digital Control

Direct Digital Control (DDC) has evolved from the growth stage of the late 1970s which were triggered as a result of the energy price hikes of 1973 and 1977. Control system technology had been evolving but a number of factors combined to make computer-based control technology more viable. One of these was the decreasing cost of electronics which made control systems more affordable. At about the same time the interest in energy savings jumped and a number of incentives and tax credits became available which stimulated the market. These factors resulted in a demand for technology that would allow building owners to save energy. These newly developed systems came to be known as Energy Management and Control Systems (EMCS).

The main computer in use at that time was the minicomputer, or central processing units (CPUs). These systems utilized energy saving features for optimizing equipment operation, offsetting electrical demand and the shutdown of equipment when not in use.

Next in this evolution was the application of Direct Digital Control. Although this technology was used in industrial process control, and even for some building applications as long ago as the 1950s, it was not until much later that it became an acceptable technique for HVAC systems.

DDC refers to a closed loop control process that is implemented by a digital computer. Closed loop control implies that a condition is controlled by sensing the status of that condition, taking control action to ensure that the condition remains in the desired range and then monitoring that condition to evaluate if the control action was successful (Figure 2-2).

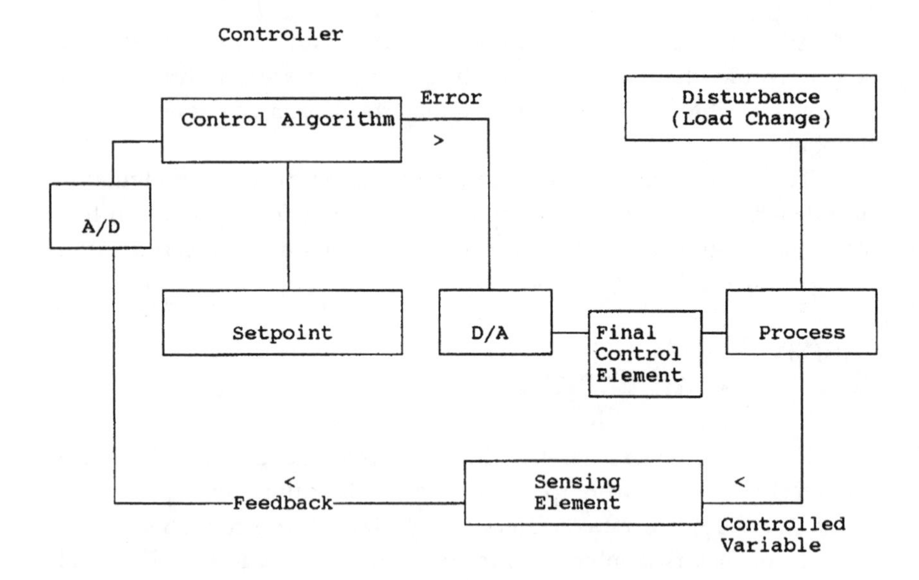

Figure 2-2. Digital control closed-loop system.

Proportional Zone Control

Consider proportional zone temperature control. The zone temperature is sensed and compared to a setpoint. If the temperature is not at the setpoint, a control action is taken to add heat or cooling to the zone. Then, the temperature is sensed again.

A thermostat with electronic or digital circuitry provides the necessary interface for digital control. It may be a microprocessor-based device that implements a sophisticated control loop, or sequence, and is capable of communications over a local area control network (LACN).

The control sequences can expand on the basic proportional temperature control and include integral derivative control. In this case, the integral is used to calculate the amount that the temperature is from the setpoint. The control action is limited to avoid overshooting the setpoint and the oscillations that cause delays in control response. These delays often occur with proportional control.

Derivative control is often used with dynamic applications such as pressure control. Derivative control is used to measure the change of speed in the controlled condition and it adjusts the action of the control algorithm to respond to this change.

The use of a Proportional, Integral and Derivative (PID) control loop allows the control variable to be accurately maintained at the desired levels with very little deviation. Other combined sequences like PID can be used to integrate the control of several pieces of HVAC equipment to provide more efficient and seamless operation. Combining this type of more accurate control with networking is an important advance for building control.

Energy Management Trends

By the mid-1980s there was no shortage of oil, and the absence of a national energy policy resulted in a decline in the demand for energy management systems. The slow but continuing growth of these systems led to an awareness of the many benefits of computerized control. Real energy cost reductions were noticed as well as other benefits of better control. These benefits include longer equipment life, more effective comfort levels and expanded building information.

The communication features available through these systems allowed improved building management and quicker response times for problem resolution. Intelligent response is a function of system communication and provides the ability to remotely diagnose a problem.

These features go beyond energy management and can be extremely desirable in rented properties. Tenants may begin looking for another office space if space comfort problems are not quickly solved.

In the future the HVAC controls industry will be driven by the following factors:

- higher energy costs and potential energy crises,
- growth in the use of Demand Side Management,
- lower cost systems and the cost-effective replacement of conventional controls,
- growth trends in performance contracting,
- growing requirements of indoor air quality and related environmental requirement,
- applications of intelligent buildings and the control integration that they utilize, and
- technology advancements including microprocessors, software, electronics and communications.

Distributed Technology

Distributed technology can be viewed as a group of control related devices and a common communication network. Building Automation System (BAS) and Building Control System (BCS) are used interchangeably. Both of these terms refer to equipment that provides building-wide control, and integrates that control through distributed devices. Distributed control devices rely on networking.

Local Area Networks and Distributed Control

In a Local Area Network (LAN) a node is any device on the network including intelligent microprocessor-based controllers. A distributed control system might control HVAC equipment and other loads such as lighting. Distributed control is applied at each piece of equipment to provide an application specific control.

Closed loop control is accomplished via the monitoring of status conditions and execution of the required control actions. One architecture defines four levels: sensor/actuator, distributed controller, building-wide host and central operator interface. Other architectures combine the host and operator interface functions.

A number of products have been introduced that use communication networks, these are sometimes called sensor or field buses. This type of technology has been growing quickly.

Remote Support

Remote support can take place through a modem interface over telephone lines. This usually means that the remote interaction necessary between the central operating interface and the local system is transparent to the user at a remote site. One promising advance in this area is building-wide controllers that support plug-and-play. This feature supports an object oriented concept, and stores all critical system information at the controller level. Intelligent controllers of this type make it possible to dial into a system from a remote location, upload from the controllers and have full access to the system.

Building-wide Controller Trends

Among the trends in building-wide control has been the drive for standard communications. Early innovators that acted without general industry consensus include IBM with its Facility Automation and Communication Network (FACN) and others from the German and Canadian governments.

Another trend at every layer of the control architecture has been peer to peer control. This technique is meaningful for building-wide control since it distributes the critical network functions to multiple controllers. The concept improves the integrity and reliability of the system. Although this type of controller can be complex, the cost premium for peer control is not significant due to the advances in electronic and microprocessor-based technology that result in greater functionality at less cost. Microprocessor technology is at the heart of building-wide and individual control advances.

Another related building-wide controller trend is integration at the functional level. There has been a movement toward integrated control between systems with different functions such as security and building control systems.

Other trends in building-wide control include easier to use programming and monitoring systems and PC-based central operator interfaces with simple user interfaces.

VAV Distributed Control

Most Variable Air Volume boxes include distributed controls. Peer controllers can be used for continuously interrogating the network for sequences such as morning warm-up. This feature would have been centralized in older systems. A single condition such as outside air temperature might have been monitored, and the building-wide device would make a decision on start time based on this data and a stored sequence. When start up was required, that controller would signal the start of the sequence. With integrated control of this type, each controller can make independent decisions based on building-wide data as well as local controller data. This results in more reliable and effective control systems.

Software Trends

Building system software is available from multiple sources today. Much of how this software should work is based upon the operation of such software as Microsoft Windows. This creates a demand for software that has a Windows look and feel for programming and monitoring. This software must be mouse driven with a point and click graphical interface. Beyond ease of use, systems are in demand that will do alarm dial outs to pagers and telephones with voice synthesis. Expanded PC-based operator interfaces may even include features like preventive maintenance and on-line diagnostics.

Nonvolatile memory storage to avoid program loss has evolved from the early minicomputer years. The technology continues to improve. The early minicomputer approach used Uninterruptible Power Supplies (UPS) to eliminate the effects of power failures. This approach was costly, but most computer hardware was costly then.

The first microprocessor-based systems used battery backup with Random Access Memory (RAM). This was as effective as the battery. Then, Electrically Erasable Programmable Read Only Memory (EEPROM) chips were used. These were nonvolatile chips that did not require either a battery on power to hold memory. This was more effective and cost effective, but the limited read/write capacity of the chip become a problem as programs and data increased. The more recent controllers use flash memory. This is a type of nonvolatile memory and has no limitation on read/write capability.

Control Trends

General-purpose controllers tend to use the same hardware, and operate at the same level of architecture, but they implement different control sequences. These sequences include such expanded applications as start/stop of non-HVAC loads and the on/off control of lighting and other electrical equipment.

In these applications there are greater requirements for control integration due to the distributed nature of the control system. At this level, as well as the building-wide level, peer-to-peer controllers are most common.

General-purpose controllers provide full local control requirements and can integrate with both the building-wide controller and the appropriate zone level controllers to provide building-wide functions.

Equipment level applications are energy intensive and include air handlers, chillers and boilers. The characteristics of the control include data point monitoring and multiple control sequences such as reset and warm-up.

Control Loops

Building-wide control systems depend on control loops to regulate the different parts of the control hardware. These can be categorized into two types, closed and open loops.

An open loop uses a controller that starts and stops the HVAC equipment at given times. There is no feedback to the controller about temperature or other controlled variables.

In a closed loop the controller responds to continuous environmental data and compares them to a programmed environmental setpoint (temperature, humidity, air quality).

In a closed-loop system, the sensing element provides feedback information to the controller. These data are compared by the controller to the setpoint that has been adjusted by the user.

The controller sends a corrective signal to the controlled hardware. The sensing element continuously monitors the environment and outputs new data to the controller. The controller monitors the sensors and activates or controls the devices to compensate for any needed environmental adjustments.

Digital Logic

The logic circuits in digital computers think and respond in terms of on and off logic. This may be a yes or no, one or zero, open or closed or true or false logic state. This binary logic is converted into electrical signals which are defined as a true or false digital signal.

The digital signal in a HVAC relay is represented by a set of open or closed contacts. The controller may interpret an input device signal as a request or non-request for heating.

A digital signal on the output of a microprocessor usually activates a contactor (relay) to either open or close. The system can use this open or closed contact to energize or de-energize an HVAC output device with an AC or DC voltage. The digital signal voltage levels may also be used directly instead of interfacing with open or closed relay contacts.

Analog Signals

An analog signal has a continuous range of values. A standard thermostat sends out an electrical signal that corresponds to temperature values. The thermostat signal is analog because it can generate an electrical quantity that can have any value between the temperature range.

In a digital system, the controller receives analog signals from sensors, and converts them to digital. The controller processes the signal and may convert it back to analog. This signal conversion is needed for HVAC equipment that requires analog signals.

Using sensor and setpoint data, the digital controller performs mathematical computations from a software program to manage HVAC equipment. The results are digital signals generated to control on/off hardware.

A/D Conversion

A digital controller may receive analog input from a sensor. Since these data are in analog form, an analog to digital converter (A/D converter) is required.

When the controller completes processing, a digital signal is transmitted. A digital to analog converter (D/A converter) may be used to convert the signal for output to the HVAC equipment. Digital controllers sample data at set time intervals rather than reading it continuously.

Digital controllers work with a set of measured control variables and a set of control algorithms. The control variables may include temperature, humidity, and air quality settings.

Control Characteristics

A digital control loop can work much more efficiently than control using pneumatic or electro-mechanical methods. The computer's operating speed allows it to sample multiple sensor devices in milliseconds. This sampling speed enables it to update many control loops quickly.

The digital software is made up of a series of command sequences that consist of computer instructions. These instructions make up the steps of the operating sequence or algorithm. Running the program translates these steps into a series of symbolic commands that the computer understands.

The algorithm or sequence of instructions may process the feedback data from the sensors. These sequences allow controllers to perform the sophisticated calculations that may be needed for efficient control.

As the controller receives data from sensors, it performs computations based on its programmed sequences. The results of this processing is sent out as corrective signals to the HVAC equipment.

Scan Rate

The power of the microprocessor used determines the execution time and scan rate. Scan rate is the process required to scan a point in the system and update the control parameters based on the current status. Proper control depends on the availability of timely information from inputs. These inputs are the source for temperature, air pressure and other HVAC measurement.

Execution time is another component in ensuring effective control. The basic requirement of closed loop control is to execute a command to outputs controlling equipment based on some measured variable. The time that it takes to do this relates directly to the microprocessor execution time.

The trend for HVAC control is for devices from process control and other automation applications to be integrated into the building industry. An example of this is the 32-bit microprocessors that have been introduced into some digital controllers.

Some equipment must be able to operate under outdoor temperature conditions. An example of this is a VAV roof top air handler. A single enclosure containing the HVAC equipment and controls is provided in one package that is mounted on the roof. The package is exposed to the full extreme of outdoor temperature and humidity and may be installed anywhere in the world.

Feedback Sensors

There are two basic types of input sensors: analog and digital. The simplest form of digital input information is an on or off condition, possibly monitored through a single dry contact. The contact allows the controller to identify a true or false condition. This can be used to identify such conditions as fan or compressor failure. With analog sensors, analog to digital conversion is needed to allow the computer to interpret the data.

Sensor trends that have been growing include the sensor bus and wireless sensors. These are related concepts since they both rely to some degree on communication.

Both of these trends are driven by the desire to reduce cost. The sensor bus brings networking communication technology to input devices. The concept is to reduce both the cost of both the labor and wire used to install sensors. This would allow greater ease and flexibility of installation as well as cost savings. It would also simplify relocating sensors that are not providing the right information on actual zone conditions.

About a third of all zone sensors are installed in the wrong location. Yet most are never moved due to the labor and difficulty involved in moving wires through walls to move the sensor. A sensor bus or wireless sensors would simplify the installation and relocation process. In a sensor bus, all sensors would be on a communication network and any controller would request these points.

Wireless sensors are another concept. Wireless sensors have been introduced that use either infrared or radio communication. The cost savings are even greater than with the sensor bus, since wiring can be eliminated all together. It is easy to move sensors, replace failed sensors and redefine a sensor to any given controller or loop.

Some wireless sensors require a clear line of sight for the accurate transmission of sensor data. With any type of radio based sensor or network interference is important. Interference that changes or interrupts communication of the current sensor information cannot be allowed to affect the control loop.

Software/Firmware

Firmware is a term used to identify anything that the average user or technician cannot change. As programmed logic, Firmware is resident in the control system. In a simple controller chip, Firmware may contain the Operating System (OS). It could also include communication protocols, data libraries and special parameters.

The OS provides the basic requirements for computing. It manages all input and output functions necessary for computing. A hierarchical structure of functions is normally used to manage the flow of information and control the activity throughout the processor.

In the past, controller manufacturers would develop proprietary operating systems and communication protocols. This meant there were no standards for communication.

Using a standard chip set for both operating system and communication is being done for bus standards such as Lonworks from Echelon. Lonworks provides a complete set of system tools.

Another trend is the use of Microsoft Windows as the basis for PC interface software. This gives everyone using the system access to compatible software and offers many tools to improve productivity.

Application specific device parameters might include information such as the number of stages of cooling in an air handler. These parameters are dynamic and allow the system to be fine tuned to the control application.

Data Trending

This involves collecting information over some time period for HVAC applications. This information makes it possible to analyze equipment operations and can be used to fine tune control sequences and identify mechanical problems. Trending provides the data needed to increase the effectiveness of HVAC control.

Setpoints

Setpoints are more dynamic than parameters since they may be changed on a day to day basis. Setpoints define the control conditions that are to be maintained for a given application. The changing of setpoint software files may be done from a central interface, hand-held terminal or even a zone sensor. Some zone sensors use a thumblewheel potentiometer or digital readout to allow a setpoint change.

The modification of a setpoint may counteract the control program. An unreasonable change to a cooling setpoint in a 65°F zone, could result in wasted energy and a higher cost of operations. A useful feature applied with most setpoints is to have a parameter that limits the amount of change allowed for the setpoint.

Another use of setpoints for equipment and general-purpose control is alarming. First, the setpoint will be established for a warning or critical alarm. These setpoints may initiate a local notification and perhaps a telephone dial out.

Alarm setpoints can be used to program different alarm conditions for the time of day. This can be done to eliminate most nuisance alarms.

Schedules

Schedules are a type of controller software that is essential to HVAC control. A fundamental requirement for cost reduction is reducing the number of hours of operation. In addition to reducing energy cost, this also extends the life of equipment since there are fewer run hours.

Schedules make savings possible in most equipment control applications, excluding hospitals and some specialized facilities. There are usually control sequences for both occupied and unoccupied modes. The schedule defines when each mode begins and ends. The setpoints are keyed to these schedules that customize the desired conditions to the time of day. This could be used from a chiller or unit ventilator application.

Scheduling is usually integrated with the occupied and unoccupied modes and time of day based alarms. Another example of integrating scheduling with other sequences is demand limiting with time of peak and time of use rates. Electric utility rates will vary, but it is common to charge more for electricity that is consumed during peak times of the day or season. The scheduling functions are integrated with the demand schedules to identify the beginning and end of such periods and to change the setpoints accordingly. In certain types of buildings such as schools and churches, schedules change daily and seasonally so quick access is important. A variable air volume (VAV) air handling unit (AHU) might implement control sequences with the priority of Table 2-2.

Table 2-2. Variable Air Volume Air Handler Operations

Priority Control
- PID closed-loop control of temperature and pressure (includes occupancy and optimization)
- Alarming

Secondary Control
- Morning warm-up
- Reset
- Return fan control
- Space pressurization
- After hours override—tenant billing
- Trending
- Safeties and diagnostics
- Preventive maintenance

Integrated Control
- Indoor air quality
- Fire/smoke evacuation
- Demand limit control
- System data

Control of Temperature and Pressure

VAV equipment must provide discharge air temperature control by sensing temperature and modulating an outside air damper for ventilation. Free cooling, along with heating and cooling equipment may also be involved.

The controlled parameters are supply air temperature and duct static pressure. Another control component is alarming. The temperature control sequence may use several loops for the overall sequence of operations. One loop or control might be used to make control decisions on the mode of operation, heat or cool. Two additional control loops, one for heating and one for cooling, then determine the amount of heating or cooling to provide. This amount is based on staging or the modulating of independent capacities, but it may also include an outdoor air economizer sequence.

Ventilation is a primary consideration and the control will ensure a minimum percent of air flow in fresh air. The outdoor air ventilation and economizer sequences should be coordinated to mix the return air from the building with fresh air from the outdoors.

Damper Control

In the larger VAV air handlers a set of sequences is used to modulate the outdoor and return air dampers to maintain a mixed air temperature setpoint. In most VAV air handler applications the control of outdoor air dampers is integrated with the cooling sequence. This sequence provides a higher percentage of fresh air based on another input, the outside air temperature. An outdoor temperature below 50°F, or higher depending on the climate, may be possible to satisfy the buildings full need for cooling through the economizer.

Economizers are usually sized to allow 100% fresh air under optimum conditions. Even more air may be used in very humid climates. The economizer may be required to consider enthalpy, rather than temperature alone, to make a decision about modulating the damper to provide free cooling.

Duct pressure control is another major segment of the control sequence. This application should improve efficiency and comfort by varying the air flow to the space based upon demand. An increase in demand at the zone may cause a damper to open thus changing the duct static pressure.

An increase in demand from multiple zones will drop the static pressure below a setpoint and require that the VAV air handler fan volume be increased. This increase is accomplished with a control loop that monitors the duct pressure and modulates fan motor speed or dampers to increase the air flow.

The HVAC control should integrate temperature and pressure control. These sequences are coordinated to ensure that a constant temperature of air is delivered with sufficient volume to meet the space conditioning needs.

Occupied and unoccupied mode control along with optimization will ensure building comfort at the lowest possible operating cost. Time of day schedules and optimized start/stop should be used in all buildings that are unoccupied for any portion of the day. There may be some exceptions such as large computer facilities.

Control Sequences

Morning warm-up is used with VAV air handlers that have a heating capacity. Most VAV air handlers are cooling only, however, in some cases the design or the climate are such that heat is used in the air handler. The greatest building-wide demand for heat during the day is in the warm-up mode prior to occupancy. Since the building temperature may be 10 to 15 degrees F below the setpoint, a great deal of energy is used to warm-up the building.

Warm-up is done with the dampers full open and the setpoint in the air handler is modified to a heat setpoint, which is usually above the standard operating setpoint. This allows heat to be quickly driven to the zone level for building warm-up.

A reset functions to change the control setpoint, temperature, or pressure and is tripped by a change in related conditions. The discharge air temperature reset for a VAV air handler might depend on the space temperature and act to reset the temperature setpoint.

The key is to identify a change in the demand due to reduced or increased load and react accordingly. The energy and cost saving comes from modifying the setpoint under milder conditions and avoiding wasted energy. This can save energy associated with the fan horsepower if the volume can be reduced due to a decrease in load. Reset is often implemented on both supply air and mixed air sequences in a variable air volume air handler.

Return Fan Control

When a return fan is installed in the VAV air handler, the controller must provide some type of fan control. The simplest technique is to interlock the fan starter with the supply fan so both always run at the same time.

A control sequence can be used to track the return fan to the supply fan. If a variable frequency drive is used to control the motor speed, there can be an energy saving in return fan horsepower. This can be integrated with space pressurization to provide indoor air quality sequences and enhanced comfort.

After Hours Control

An override function allows access to the building by a tenant with conditioning provided during these times. It can be initiated by a push button, switch or dial-up with the central operator interface or touch tone telephone interface to the controller.

There is normally a fixed duration of time that the override will last, but the amount of time that a controller is in override may also be a variable. It could be based on the time needed to log the status of various points in the system.

Modern leases may include access to the space on a 24-hour basis, but only provide space conditioning during normal working hours. Occupants who use the space after hours may be required to pay a use charge for the lighting and space conditioning. This is known as after hours tenant billing, and has become a common practice in many office buildings.

Space Pressure

A control loop that is used more often with VAV air handlers involves monitoring and controlling the space pressure. This type of control is used in environments where a negative pressure is sustained to avoid the flow of contaminated air outside the controlled space.

Maintaining a slightly negative pressure induces the flow of fresh air into the building. Maintaining neutral space pressure prevents the loss of expensive conditioned air to the outside from positive pressure. Positive pressure can result in noise around diffusers that may bother the occupants.

Alarming

One of the benefits to intelligent control is the ability to monitor a variety of alarm conditions. Most of these are simple control loops that monitor the condition of a contact closure or an analog value. Based upon a change in state of the digital point, or an analog point exceeding a set of range limits, an alarm is initiated.

Safety Sequences and Diagnostics

Safety sequences are used to produce these alarms. These control sequences are implemented to protect the HVAC equipment as well as the building and occupants. The trend is toward more safety sequences. In a VAV air handler these may include duct high pressure, compressor pressure, freeze and low discharge air temperature alarms.

Diagnostics are program sequences that are used to identify problems. They are used to interrogate the system and diagnose problems by reviewing such parameters as temperature or pressure. Diagnostics can also be self-test routines. The use of safety and diagnostic sequences can optimize controller operations and reduce downtime.

Diagnostics and preventive maintenance are related, but preventive maintenance sequences are used to increase equipment efficiency, extend equipment life and provide operation with a minimum of downtime. They may collect data on run hours of the equipment and use alarms to flag problems that may require immediate attention. Other program sequences may be used for indoor air quality, fire/smoke evacuation and demand limit control.

Indoor Air Quality

These program sequences may introduce additional fresh air to maintain the parts per million of carbon dioxide (CO^2) or reduce the amount of fresh air introduced to the building if contaminants are detected outside the fresh air intake. A sensor may be located near the outdoor air intakes to monitor for carbon monoxide, or other contaminants. The fresh air intake will be stopped if these are detected.

Zone control may include sensors to monitor the carbon dioxide. The air handler will add fresh air as required to purge the space and reduce the concentration of CO^2. Other control functions may involve exhaust or return fans and dampers.

Fire/Smoke Evacuation

The smoke pressurization sequence may involve all VAV air handler controllers as well as the zone controllers. Providing positive pressure on the floors above and below a fire may be part of these sequences. The fire floor may be controlled to a negative pressure with all zones shut down, and return/exhaust fans at full volume.

Smoke evacuation is a sequence which enables the exhaust and return fans on all units. Another technique to use separate smoke evacuation fans that are enabled when a fire occurs. In this case all VAV air handlers are shut off to avoid introducing more air or interfering with the smoke evacuation fans.

Demand Limit Control

This control technique raises the cooling setpoint in order to reduce some stages of cooling. This is a building-wide sequence that requires equipment turn-off and avoids demand peaks. If companies know their energy profiles, how and when they consume power, they can negotiate better rates for the type and amount of power they need.

Power Measurement

Intelligent devices and systems can monitor, measure, protect, coordinate, and control how power is distributed. Power and control monitoring systems use meters, protective relays, circuit break trip units, and motor starters. They can communicate information over an Ethernet network to a central location for remote monitoring, alarming, trending, and control.

Power-monitoring software can be used to analyze energy use and power quality. It can identify load profiles to help rate negotiation. Load shaping involves the prediction of demand excursions for shedding loads or starting generators to avoid setting new peaks. It can also be used to predict maintenance requirements or provide energy bills for individual departments or areas for actual power use.

Intelligent metering and monitoring systems offer a low-cost method for quick energy saving practices. The Cutler-Hammer plant in

Asheville, NC, installed a power management system in early 1997. Energy bills were running close to $45,000 a month. After 6 months of installation, the plant energy saving amounted to $40,000.

The system allowed plant engineers to identify wasteful procedures, shift loads to level demand and perform preventive maintenance. There was better control of area lights at off hours and large electric ovens were used only during the late shifts when total energy demand was lighter.

Maintenance technicians were able to pinpoint abnormal conditions with monitoring screens and then service the equipment before it broke down. Total return on investment was predicted to be less than two years.

Control and Power Management

Linking power management systems to control systems allows the power information to flow from both systems. Load profiles can be developed to discover any energy inefficiencies. Energy scheduling can be done to find the optimum energy schedule for new products or processes. Real-time utility pricing means that production schedule energy requirements need to be compared with energy rate schedules for optimum energy benefits.

The new energy supply market requires more companies to give back energy capacity during peak energy use times by scheduling lower-energy production. This often results in significant savings.

Power Quality

Power quality may have significant effects on product quality. Voltage sags or swells may result in product defects. A power management system can be linked with a quality management system to help identify power-quality-caused problems. This can reduce reject rates and improve productivity.

Monitoring power consumption can also reduce maintenance costs. Monitoring the energy a motor is drawing can show you when a tool begins to get dull and produce bad parts or when a motor is overheating. Replacements can be made as preventive actions without costly downtimes or producing defective products.

Energy controls are only part of an environmentally conscious policy. Modern civilization is electricity based, but over the past 20 years most of the energy savings has mainly been in heating.

Energy Management Devices

Energy management devices are used to regulate the on and off times of selected loads, such as fans, heaters, and motors. They are used in a building in order to reduce electrical demand (kilowatts) and to regulate energy consumption (kilowatt hours). Energy management devices can be electromechanical, electronic, or computer based.

The operation of one or more loads is interrupted by the energy management system based on control algorithms and building-operating parameters, such as temperatures, air flow, or occupancy. The savings in electrical energy use and cost range from 0 to 50% or more.

Electrical Demand

Electrical demand is defined as the average load connected by a user to an electrical generating system. It is measured over a short, fixed period of time, usually 15 to 30 minutes. The electrical demand is measured in kilowatts and recorded by the generating company meter for each measurement period during the billing month. The highest recorded electrical demand during the month is used to determine the cost of each kilowatt hour (kWh) of energy consumed.

Demand Load Shedding

Energy management devices, such as load shedders, reduce the demand (average load) in critical demand periods by interrupting electrical service to motors, heaters, and other loads for short periods. Since the load which has been turned off would normally have been operating continuously, the overall effect is to reduce the average load or demand for that period of time (Figure 2-3). The instantaneous load when the load is operating remains the same. If the period involved has the highest monthly demand, considerable savings are possible in rate reductions. In periods other than the highest demand period, some energy is still saved.

Before the era of high energy costs, load shedding was used mainly to avoid demand cost penalties. Now, it is used to limit energy consumption, by cycling loads on and off for brief periods, as well as to reduce demand.

Other methods used to limit energy use include computer optimization of start times, setpoints, and other operating parameters based on weather, temperatures, or occupancy.

Meter Connection

Energy management devices base their control actions on the demand level in each demand period. They get two signals from the electrical utility meter (Figure 2-4). These signals are usually momentary contact closures of about 500-ms duration which indicate the beginning of each demand period or the consumption of one kWh of electrical energy.

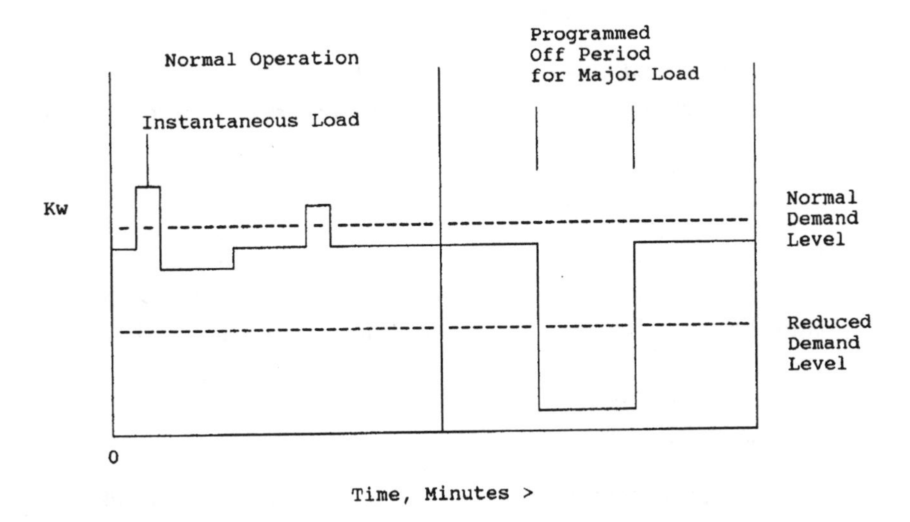

Figure 2-3. Reduced demand from cycling a major load off for a brief time period.

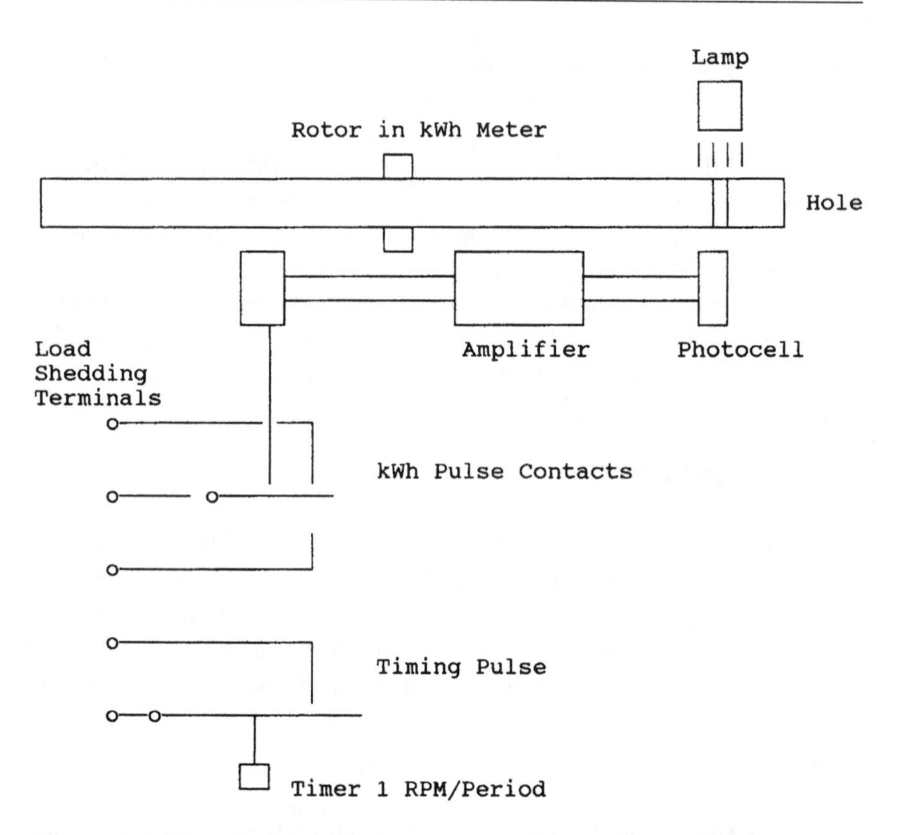

Figure 2-4. Electrical meter connections. The timing pulse marks the end of each demand period. The kWh pulse occurs at each rotation of the meter rotor.

Meters can be modified by the utility company for about $500 to $1000. External clock mechanisms or timing signals can also be used. The energy (kWh) consumption pulse value is converted to a unit of demand (kW) by dividing it by the demand period length in hours.

$$\frac{\text{kWh per Pulse}}{\text{Demand Period in Hours}} = \text{kW (Demand)}$$

The simplest power control device is a timer or time clock. An electromechanical timer or clock has a small electrical motor and an arrangement of cams and switches. These switches allow several loads to be operated in a cyclical mode by turning them on or off at preset times. Electronic timers or clocks use semiconductor timer chips and switches to perform the same functions. Popular chips like the 555 and 556 can perform these functions and cost less than 50 cents.

Applications range from pool heaters to advertising signs and outdoor display lighting. They have also been adapted to control electric energy use and to limit demand. The timer is set to turn a load off and on for brief time intervals during each demand period. This has the overall effect of limiting energy consumption and reducing demand. The control is limited to those devices which can be cycled continuously.

Electronic Demand Limiting

Other types of electronic demand limiting devices can be used to monitor and measure the actual demand and provide control actions to limit the operation of attached devices when the measured demand reaches a preset value. These devices need to receive two signals from the utility meter, the kilowatt hour (kWh) or demand pulse, which indicates the units of electrical energy consumed and a timing pulse, which indicates the end of one demand pulse and the start of the next one.

The demand pulse is used to increment an electronic counter. As these counts accumulate within a demand period, they are compared to a target value which corresponds to the maximum desired demand. When the value accumulated in the counter exceeds the target value, the attached load is switched off.

If the demand still exceeds the demand target, a second load is turned off. This is done until all available loads are turned off or until the demand no longer exceeds the target.

Once it is turned off, the load stays off until the end of the demand period. A timing or reset pulse from the meter restores the counter to zero and turns the loads back on again. Some load shedders use a demand target that is not fixed but increases at a steady rate. Other devices allow the off-on setpoints to be adjusted independently for individual loads.

Computer-based Systems

Using computer technology in the regulation of electrical energy allows a wide range of control options. The computer's memory and decision-making circuits are combined with a customized program to maximize demand and energy savings with a minimum impact on operating conditions. The computer receives inputs from the electric meter and from sensors monitoring the critical loads in the building or facility.

Loads may be cycled based on the following criteria:

- maximum demand target,
- time of day and day or week,
- rate of demand increase,
- heating and cooling temperatures,
- pressures,
- fuel flow and rates,
- occupancy schedules,
- inside and outside temperatures,
- humidity,
- wind direction and velocity and
- combinations of the above factors.

The durations are variable and can be changed automatically according to these parameters. A secondary demand target may be used and if the primary demand target is exceeded other control actions are applied.

The demand target can float upward and downward according to demand rates, time of day, and other parameters. Different targets and control strategies can be applied to different sections of buildings and plants. Reports and graphs showing the consumption by demand period and by day, week, or month are provided.

These systems can also be applied to manage the energy consumed by air conditioning systems. Intake and exhaust dampers are controlled on the basis of air temperatures, so that the mix of air requiring the least energy is obtained at all times. The start-up and shutdown of air conditioning, heating, and lighting systems can be regulated according to inside and outside temperatures as well as occupancy. The goal is to produce the conditions which consume the least energy.

Energy Efficient Lighting

Buildings need new ways to reduce energy costs while maintaining quality lighting for different facilities. A large portion of facility utility costs are devoted to lighting. Recent light innovations and energy conservation programs are helping to facilitate this task. Technological advances in lamps, ballasts, and lighting controls, such as improved metal halide systems with pulse start ballasts are transforming this sector of energy saving.

Lighting upgrades save energy dollars. Many facilities suffer from outdated lighting components that are nearing the end of their life. Ballasts may be 20 years old and their life expectancy was 15 years.

More efficient buildings are the goal of energy conservation organizations, such as the Energy Efficient Lighting Association (EELA). It specializes in effective retrofitting and the latest lighting technology. The EELA was launched in 1997 and is based in Princeton Junction, NJ. EELA sponsors conferences and is supported by lighting service companies.

EELA helps facilities managers achieve energy-saving goals with an array of lighting products. By funding lighting retrofits, energy services companies (ESCOs) also help to achieve energy reduction.

In a study for a large hospital, EELA recommended a lighting retrofit with electronic ballasts and energy-efficient lamps. A local utility company offered a rebate that covered a portion of the costs. A few years later, the hospital found that it had saved several hundred thousand dollars in energy costs.

Companies such as Prescolite are incorporating feedback from installers, end-users, and design professionals in energy-efficient light product design. One example is energy-saving universal voltage ballasts that accept multiple wattages and lamp types. These are beneficial to facilities managers and installers.

Other developments for innovative solutions include advances in dimming ballasts for compact fluorescent lamps with direct digital dimming and end-of-life circuitry. Microprocessor control technology in ballasts is also reducing energy costs and providing flexibility for end-users. This is the future in dimming control technology.

Buildings can be improved with updated ballast technology and group relamping. Group relamping is the opposite of spot relamping and it saves on maintenance, as well as utility costs. Many facilities attempt to remedy their situations with piecemeal lighting solutions. Retrofitting provides an average energy savings of 50 cents for each square foot of building space. The average payback is 2-1/2 years. Partnering with energy service companies can also help with energy audits and funding lighting upgrades.

Demand Side Limiting

Most utility programs for energy conservation in recent years involve demand-side management (DSM). These programs impact how customers use electricity. The concept has two parts. The first tactic is to even out the demand for electricity so that existing power stations are operating at efficient capacities throughout any 24-hour day rather than rocketing up during business hours and dropping down during the night. The other part of DSM is to deter the need for new electricity capacities.

DSM refers to energy and load-shape modifying activities that are implemented in response to utility-administered programs. It does not refer to energy and load-shape changes arising from normal actions of the marketplace or from government-mandated energy-efficiency standards.

During the late 1980s, utilities began offering commercial rebate programs for DSM. Some utilities pay 30 to 50% of the installed cost, while others base their rebate programs on the peak-kilowatt-demand savings achieved by new equipment. DSM programs consist of planning and monitoring activities which are designed to encourage consumers to modify their level and pattern of electricity usage.

In the past, the primary object of most DSM programs was to promote cost-effective energy usage to help defer the need for new sources of power, including generating facilities, power purchases, and transmission and distribution capacity additions. Due to the changes that are occurring within the industry, electric utilities are also using DSM as a way to enhance customer service.

DSM involves peak clipping, strategic conservation, valley filling, load shifting, strategic load growth and flexible load shaping. It may include interruptible services or curtailment of services for specified time periods for commercial customers. Peak clipping refers to reducing the customer demand during peak electricity use periods. This is done by using some form of energy management system.

Strategic conservation is often rewarded by utility rebate programs. It can include building energy audits, weatherization, high-efficiency motors, Energy Management, DDC systems and HVAC systems and equipment.

Valley filling increases the electricity demand during off-peak periods, which allows the utility to use its power generating equipment more effectively. Load shifting is like valley filling, since it uses power during off-peak periods. Both valley filling and load shifting programs often involve thermal storage systems.

Strategic load growth is a related DSM program that encourages demand during certain seasons or times of the day. Flexible load shaping modifies the load according to operating needs and can result in interruptible or curtailment rates for customers.

Consolidated Edison's program involves organizations that can reduce their summer electricity bills without buying new equipment. During the summer months, these customers agree to reduce electric demand by at least 200 kilowatts on demand. More than 100 organizations were involved in this program. Duquesne Light Company in Pittsburgh and Georgia Power also have interruptible economic development rates that operate similarly.

Con Edison also offers programs with energy audits and rebates for efficient lighting, steam air conditioning, gas air conditioning, high-efficiency electric air conditioning, cool storage and high-efficiency motors.

Houston Lighting & Power (HL&P) has a program to encourage the use of cool storage technology. It provides building owners with a $300 cash incentive for each kilowatt reduction in peak demand. There is also a cool storage billing rate, which defines the on-peak demand as noon to 7 p.m. Monday to Friday throughout the year. Many buildings have increased in value and marketability as a result of these cool storage programs. In the Dallas/Fort Worth area, Texas Utilities has more than 135 cool storage systems in operation.

Kraft General Foods and Boston Edison have an Energy-Efficiency Partnership that has reduced the ice cream manufacturer's cost dramatically. This project decreased the cost of producing ice cream by one third. The ice cream manufacturer was able to upgrade most of its electrical energy-consuming capital equipment and obtain substantial rebates for the energy saved. The rebates returned more than 85% of a $3 million investment. This included refrigeration and defrosting equipment, lighting installation and monitoring equipment.

Georgia Power has its Good Cents building program for commercial customers with HVAC and lighting rebates, along with energy audits. Georgia Power has also developed an indoor lighting efficiency program.

Besides rebates there are low- or no-interest equipment loans, financing, leasing and installation assistance and assured payback programs. Wisconsin Electric Power Company offers rebates of up to 50% of the project cost and loans with multiple rates and terms for 3 to 7 years. These programs are available to building owners and managers who install energy-efficient lighting, HVAC systems, window glazing, high-efficiency motors or building automation systems.

Direct Digital Control Programs

About a third of all utilities offer rebates for controls. These controls include time clocks, lighting controls and Energy Management Systems. An Electric Power Research Institute (EPRI) survey found that the savings from these systems were 15% of the building energy use. Paybacks ranged between 1.5 and 3 years for control system projects including DDC systems.

Commonwealth Edison Company in Chicago offers its Least Cost Planning load reduction program. In this program, businesses cooperate to curtail or reduce their electricity consumption to prescribed limits when the utility requests it. They are compensated with a special electricity rate that is performance-based. The worst performance during any curtailment period becomes the base for electricity charges.

DSM Trends

Increasing competition among electric utilities may curtail DSM programs in the United States. A number of utilities are cutting back on DSM program budgets while others are shifting the focus of their DSM programs to minimize the impact on electric rates.

According to the Edison Electric Institute (EEI), DSM programs grew from 134 in 1977 to nearly 1,300 by 1992. These DSM programs deferred more than 21,000 megawatts (MWs) by 1992.

In 1997, about 1,000 electric utilities had DSM programs. A little more than half of these are classified as large and the rest are classified as small utilities. Large utilities are those that produce more than 120,000 megawatt-hours. This group of larger utilities account for about 90% of the total retail sales of electricity in the United States.

Energy savings for these large electric utilities decreased to 56,406 million kilowatt-hours (kWh) in 1997. This was 5,346 million kWh less than in 1996. These energy savings represent 1.8% of annual electric sales of 3,140 billion kWh to ultimate consumers in 1997.

Actual peak load reductions for large utilities was 15% lower in 1997 at 25,284 megawatts, than in 1996. DSM costs decreased from $1.9 billion in 1996 to $1.6 billion in 1997. This was the fourth consecutive year that DSM costs have decreased from a high of $2.7 billion in 1993.

Rebate and incentive programs are being modified to reduce utility costs, increase participant contributions and increase program cost effectiveness. Rebates are being used more sparingly and tend to focus on one-time energy efficiency opportunities. These programs are skewed to customer classes that are underserved by other energy efficiency programs and energy efficiency measures that are difficult to promote through financing, education and standards programs.

Building and Equipment Standards

Utilities are supporting adoption and implementation of stricter building codes and equipment efficiency standards. Appliance and equipment efficiency standards are having a notable impact on electricity demand in the United States. Standards have lowered national electricity use by 3%. Some energy efficiency measures, such as power-managed personal computers, have been widely adopted without financial incentives or much utility involvement.

Intelligent Buildings

The increasing acceptance of energy management systems for building management applications has been pushed by federal mandates. Energy saving systems integrate the operation and management of heating, ventilation, air conditioning, security, light and fire safety systems to reduce energy costs and minimize carbon dioxide emission of commercial buildings. The weak link in most older systems is the dependence on a human operator.

The future vision is a building that almost runs itself, from adjusting HVAC loads to dimming the lights. Many existing buildings should include such integrated energy controls systems by 2005. New constructions will be designed to utilize practical, integrated building controls.

Energy efficiency is part of an overall goal to reduce energy use and the carbon dioxide emissions that increase the greenhouse effect. The result will be practical, computerized energy management systems that unify the operation and monitoring of heating, ventilation, air conditioning, security, lighting and fire safety systems. These systems will be largely self-managing, correcting changes within the building automatically and alerting building personnel when problems occur.

Most of the building control systems being designed come with built-in computer network connections. These allow the individual controls systems to communicate either directly with each other, or with a local or remote PC.

Control Trends

Many large firms such as AT&T and Boeing Aircraft have been using PC-based management and control systems. These systems are based on a multitasking PC operating system (OS) such as Microsoft Windows NT or on more powerful workstation-based systems.

AT&T has used its electronic mail system to connect about 80 buildings around the country with over 25 million square feet of space into a single PC located in Kansas City, MO. Boeing Aircraft has used a similar approach into a Windows-based system.

Computerized energy controls are the wave of the future, but a critical part is the interface between the computer and the operator. This interface is concerned with what the operator sees on the computer screen and the tools the operator has to manage the system. The interface cannot be too difficult to understand or use. More powerful software can allow operators to understand the underlying mathematical and algorithmic dynamics, and how each change affects the whole.

The controls must act as an integrated system since each change in the system can affect the whole. These systems cannot be optimized one element at a time, without taking into consideration unpredictable results.

Open Automation

Open automation systems are evolving because of advances in technology and new ways of doing business. The benefits that are gained include agility, quality, reduced downtime, and lower costs. Open systems are nonproprietary and based on standards so components from multiple vendors can interoperate in a predictable manner and be integrated easily with other systems.

A PC can be used to program control systems, run control, and provide operator interface and data analysis. Open automation systems consist of open operating platforms, interoperating components and open networks. They form the backbone of improved control in the future.

A Sense of the Future

During the last half of the 20th Century, industrial controls moved from pneumatic controls, first to vacuum tubes and then transistors and integrated circuits (ICs). Direct digital controls (DDCs) gave way to distributed control systems (DCSs) which are now implemented with networked PC-based control systems (see Table 2-3).

From the roots of contactors and motor controls came programmable logic controllers (PLCs) which offered better alternatives for control. Now, PLCs are almost considered commodity items. With the growth of Internet connectivity, visions of vast arrays of interconnected systems are possible.

New and unconventional discoveries and developments will result in changes significant as the microprocessor and the Internet. In the biotech arena, the discovery of DNA occurred and soon there will be detailed genetic and physical maps of the human genome. This should decide the complete nucleotide sequence of human DNA and localize the estimated 50,000-100,000 genes in the human body. This research is expected to be complete early in the millennium and should have far reaching effects.

Table 2-3. Major Events in Industrial Control

Early 1940s	Pneumatic Transmissions
Mid-1940s	First PID Controllers
Late 1940s	Transistor Invented
1950	Electronic Analog Control
Late 1950s	First Control Room HMI
Early 1960s	Software
	Computer Digital Control (DDC)
Late 1960s	Integrated Circuits
	First RAM Chip
Early 1970s	Fax Machine Proliferation
	Microprocessor Introduced
	PLC Ladder Logic
Mid-1970s	First PC Introduced
	Distributed Control System (DCS)
Late 1970s	Object Oriented Programming
Late 1980s	Cell Phones
	PC-based HMI Software
Early 1990s	Internet Use Explodes
	Field-based Architectures
Mid-1990s	Intelligent I/O
Late 1990s	E-commerce Starts
	Complex Adaptive Systems
Early 2000s	Third-Generation Wireless
	Human Genome Project Complete
	Direct Digital Line to Brain
Mid-2010s	Nanotechnology
	Bio-chips Introduced
Late 2010s	Moore's Law Obsolete

Biochemical Computers

Scientists are making computers out of strands of synthetic DNA rather than silicon chips. These computers perform computations simultaneously rather than one at a time, as today's machines do. This makes them potentially much faster.

The element silicon is closely identified with computers but such thinking may have to be fundamentally revised, as high-speed computation moves beyond chips and machines to include the tools of biochemistry and genetics including DNA.

DNA is present in every living organism, and the appeal of the molecule as a computing mechanism lies in its ability to store a vast amount of information in a small space. Instead of using zeroes and ones to encode information using electrical current, the memory in a DNA computer uses the DNA strands that are synthesized in a lab. Each strand contains a different sequence of the chemical bases, given the letters A, C, T, and G, that make up all DNA molecules. Scientists use different enzymes to eliminate certain strands of DNA, leaving only the strands of bases that represent correct graphic answers.

Scientists at the University of Wisconsin have performed simple calculations with strands of DNA that are attached to a gold-plated surface. Other experiments with DNA computing allow the DNA to float freely in a test tube.

The method used at the University of Wisconsin uses single strands of DNA which are anchored to an inch square, gold-coated slide. Each slide has a unique configuration of biochemicals that represents a possible answer to a problem.

The slide is washed with a series of biochemical solutions that mark and eliminate incorrect answers all at once by breaking apart the molecules. The remaining DNA strands are decoded, which represent the correct answers to the problem.

Another group at Princeton University are using RNA, a chemical sibling of DNA, to perform similar calculations. One calculation provides the answer to a chess problem known as the knight problem. This involves the positions of the knights on the chess board so that none can attack another.

Each strand of RNA represents a possible configuration of knights. Then, chemicals are used to eliminate RNA strands that depict wrong answers.

A few grams of DNA could store all the data known to exist in the world so biochemicals could eventually be the most efficient medium of storing and manipulating information. Another advantage over conventional computing is that rather than analyzing each possible answer in sequence, the DNA computer can act on the entire set of molecules simultaneously.

The first applications for DNA computing might involve the encryption of large amounts of information and embrace some combination of silicon and DNA computing.

Device Communications

For many years we have been hearing that soon everything will talk to everything else. Connection problems may be on the way out due to three new technologies: Jini, Universal Plug-and-Play (UPnP), and Bluetooth.

Jini was conceived by Sun Microsystems in 1994 as a plug-in powerline network. It allows a miniature operating system (OS) in every piece of hardware. When you plug the device in, it sends vital information to a central network registry identifying the device and its location. The computer then checks on what devices are available and downloads any information needed to make the connection. Jini lets you plug computer hardware into any network because there are no device drivers.

In its Windows 95 introduction, Plug-and-Play promised to simplify system setup and expansion. Microsoft wants to extend the concept to networked peripherals and intelligent appliances. The new generation called Universal Plug-and-Play (UPnP) is expected to automate device recognition in networks without the need for a centralized server. UPnP extends Plug-and-Play to devices you may not hook directly into the PC, but that still work on the computer network.

The Bluetooth specification hopes to unite the networks, phones and appliances. It is backed by more than 1,000 companies, including IBM and Intel. Bluetooth builds an ad hoc network when another Bluetooth device is within the range of about 33 feet. Bluetooth could be used in smart refrigerators which send food shortage lists to the phone which in turn sends them to the grocer.

Several ground rules may be important early in this new millennium. These rules or laws are often celebrated as having some control on the spread of technology. Moore's Law was formulated by Gordon Moore of Intel in the early 1970s as a way of estimating the growth of microchips. It states that chip density or processing power doubles every 18 months. This infers that as computers become faster, the price of a given level of computing power halves every 18 months.

Gilder's Law as proposed by George Gilder stipulates that the total bandwidth of communication systems triples every 12 months. New developments in communications appear to confirm that bandwidth availability will keep on expanding at a rate that supports Gilder's Law.

Metcalfe's law comes from Robert Metcalfe, the inventor of Ethernet and founder of 3COM. It states that the value of a network is proportional to the square of the number of nodes. This means that as a network grows, the value of being connected to it grows exponentially, while the cost per user remains the same or even drops.

Moore's Law applies to ICs and may no longer be applicable as IC geometry becomes about one atom thick. Other advances in microelectronics may also affect this. Lucent, among others, is conducting research that could allow the fabrication of multiple transistors vertically in silicon.

This suggests a cubed law instead of a squared law, which could provide another 20 years of doublings before the limit is reached in all three dimensions. However, other technologies, such as bio-chips and nanotechnology, will become prevalent and override the equivalent of Moore's Law in the future.

In the future, many people will use wearable computers. A fashion technology show at the Spring Internet World 2000 features the latest wireless devices for connecting to the Internet.

Smart phones are cell phones that have Internet and organizer capabilities. Some such as the NeoPoint 1000 have voice recognition while others like the Nokia 7190 use a dictionary that finishes words as you input them. SwatchTalk is the world's first watch with a built-in phone.

Intelligent I/O Appliances

In the present state-of-the-art, dumb sensors are connected to processors. The intelligence resides primarily in the central master. As the price of processing power and memory drops, embedded processors will be used in most sensors making them intelligent appliances (see Table 2-4).

The specifications of each sensor, actuator and measurement or control product including part number, installation date, characteristics, instructions and diagnostics will be embedded in the product and be accessible to the information network.

Table 2-4. Smart Appliance Characteristics

Dumb Motor Information Provided	*Smart Motor Information Provided*	*Additional Features*
Power on/off	Specifications	Speed Control
Overload Trip	Maintenance	Soft Start/stop
Single Direction	Replacement	Braking
Hard-start	Lubrication Run Status RPM Direction Torque Power Current Voltage Temperature Bearing Condition Bent Shaft Detection	Phase Correction Acceleration Control Torque Control Programmable Logic Alarm Managing Pre-failure Warnings
Interface Power	Power	Smart Control Network with Embedded Object & Intelligence

At one time, signal conditioners performed a single specific function such as converting a sensor output to a high-level, linear signal that was then sent to other instruments. With the advent of inexpensive processing, memory and communications circuitry, the signal conditioner can now perform several functions and is not limited only to single linear functions. The intelligent signal conditioner could be interrogated to find out the control state during a particular time period and investigate faults. This information can be used for analysis and diagnostics.

In the future, PCs will replace PLCs in all but the smallest applications. This is because of the higher-level programming capabilities and low-cost communications and networking capabilities of PCs. PLCs may serve as machine controls and as front-end I/O processors for device-level networks. Traditional DCS will give way to field-based intelligent I/O systems. The older centralized, hierarchical control architectures will fade because the software requirements of central command and control systems are becoming burdensome and practically unmanageable.

Complex Adaptive Systems

There will be a transition from traditional procedural controls to rule-based, hostless peer-to-peer controls systems (Figure 2-5). When each input or output point becomes intelligent and connected, control will be handled locally through intelligent peer-to-peer communications among the sensors and actuators, without intervention from an external host processor.

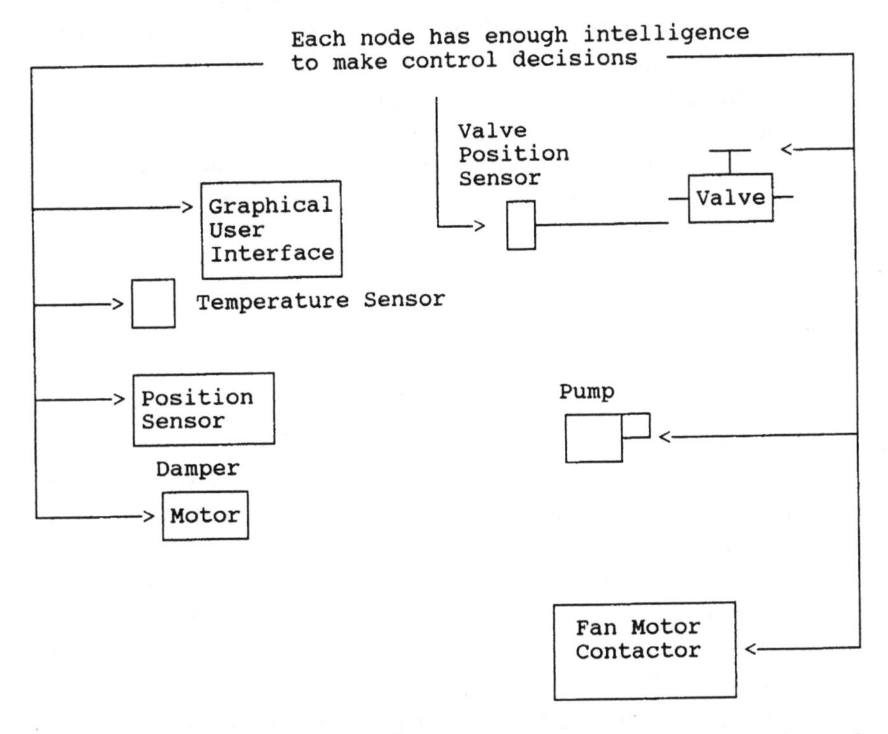

Figure 2-5. Architecture for complex adaptive systems, with I/O and intelligent peer-to-peer communications for true distributed control.

Beyond the improved effectiveness of local control, the increased use of peer-to-peer control systems brings Metcalfe's Law with its exponentially increasing effectiveness into play. This results in a different type of control which is the use of complex adaptive systems.

The operation is much different from that of conventional, deterministic, hierarchical control. Independent, intelligent, autonomous I/O and controls operate and interact using rule-based mechanisms. As the complexity of the interaction increases and changes, the system responds and adapts to the changing events and requirements. These changing events could be fuel availability and costs or seasonal variations. This is the equivalent of genetic algorithms which exclude nonproductive operations and stimulate improvements.

There will be significant advances through reduced software, faster and easier installation, more robust performance, improved flexibility and the ability to handle more parameters and parameter variations. The system will be more robust because performance is not dependent on single failure points. The failure of any single portion of the system is accommodated. Higher levels of performance are also achieved through emergent behavior and self-organizing capabilities.

These systems are rooted in the work being done on artificial life and genetic algorithms. During the next decade, the results of this research will become visible in factory automation and process control systems.

Software

In the future the notion of software as a separate and external entity will largely evaporate. Software will become a part of the product in which it inhabits similar to today's firmware. Many of the current software functions will migrate downwards into firmware and object-driven applets.

The only external software needed will be in the browser. Everything else will be Java applets or similar client-type operations triggered by the servers when connected, or by the objects that reside within the products.

Among the products that have generated much of the growth for software companies in the past were graphic human-machine interface (HMI) packages and applications that furnish display and control functions. Some of these products are an outgrowth of supervisory control and data acquisition (SCADA) systems.

Industrial Networks

In modern factory and plant environments, any islands of automation in the past are steadily melting away with the connection of essentially everything to nuclear networks.

Today, there are three levels of networking:

- enterprise networks up at the top,
- control networks in the middle level, and
- device level networks at the lowest level.

The fieldbus standard may evolve to the practical emergence of several parallel and perhaps overlapping standards. Each of these may be suited to a particular industry and/or environment. Industrial extensions of Ethernet and TCP/IP will become the defacto standards for connectivity external to the system or machine.

The future will see the proliferation of Ethernet at all levels. Its performance and cost will stamp out the need for alternatives. Low cost will be fed by acceptance and high volume in most application areas.

Wireless Connections

Worldwide connectivity is moving very quickly to link everyone and everything to the Internet. This will take place not only through high-speed DSL and cable modems, but also through third-generation wireless systems. The analog technology of first-generation cell phones slowed the adoption of advanced digital wireless technologies in the United States. Because of the entrenched infrastructure of the analog telephone system the initial growth was slow compared to Europe, which quickly jumped directly to second-generation wireless digital telephones. In the near future, advances in low cost microcircuits will allow broadband wireless connections at speeds exceeding twisted pair landlines and nearing that of cable connections.

This will have a major impact on control systems. Connecting control components conventional wire systems beyond the confines of a typical system enclosure is still a major task in many typical HVAC applications.

Future controls will be accessible through portable wireless computers carried by individual maintenance people, much as cell phones are carried today. These portable wireless PDAs (personal digital assistants) will be connected via the Web to everything needed to read the local objects. These objects will include the specifications for each control component as well as maintenance, troubleshooting and repair documentation.

Third-generation wireless connectivity will include the new local area network technology called Bluetooth that allows I/O data transfers at 1 megabyte per second within a 10-meter radius.

One inhibition to wireless operation is the fear of interference and tampering. This can be eliminated through modern encryption techniques, which are already being used in most banking, stock market and other high-value applications. The adoption of these security methods in industrial automation should be relatively straightforward and complete the cycle of integration.

Diagnostics

Intelligence allows diagnostics which can conduct not only in post failure, but it can also be predictive and advisory in nature while providing maintenance instructions.

It is extremely useful to know that a failure may occur in the near future. If a valve is sticking, it might be allowed to continue to operate with some precautions. If the valve exhibits a particular symptom, then immediate maintenance may be needed.

Predictive diagnostics allows components to be repaired or replaced during off-season downtimes, before a problematic failure occurs during HVAC operation.

This type of intelligence may reside not only in central computers, but also in an increasing number of controllers, actuators and displays. The intelligence aspect could become a part of essentially every product in the control system. Some types of failures are best diagnosed using methods that involve failure simulation and pattern recognition through the responses of intelligent I/O points in a complex adaptive system.

References

Currier, Bob, "Let's All Get Along: Your Refrigerator and Oven are About to Get into a Heated Debate," PC Computing, Vol. 12 No 1, January 2000, p. 52.

Joch, Alan, "Plug-and-Play for Manufacturers," Byte, Vol. 21 No. 9, September 1997, pp. 92-100.

Liptak, Bela A., Editor-in-chief, *Instrument Engineers' Handbook*, Revised Edition, Chilton Book Company: Radnor, PA, 1982.

McGowan, CEM, John J., *Direct Digital Control: A Guide to Distributed Building Automation*, Fairmont Press, Inc.: Lilburn, GA, 1995.

MacNeil, John S., "The Wet and Wild Future of Computers," *U.S. News & World Report*, Volume 128 No. 6, February 14, 2000, p. 52.

"Network Solutions," *Intech*, Vol. 46 No. 8, August 1999, pp. 34-35.

Pinto, Jim, "Instrumentation and Control: On the Frontiers of a New Millennium," *Instrumentation and Control Systems*, Vol. 73 No. 1, January 2000, pp. 29-39.

Priscaro, Frank, "What's Next," *MC Technology Marketing Intelligence*, Vol. XX No. 2, February 2000, p. 64.

Pulaski, Frank, "Power Down the Plant Costs," *Industrial Computing*, Vol. 18 No. 7, June 1999, pp. 13-15.

Reinback, Andrew, "The Buzz About Energy Controls," *Buildings*, Vol. 86 No 10, October 1992, pp. 50-54.

Internet: www.eiadoe.gov/cneafelectricity/dsm/dsm_sum.html, "Electric Utility Demand-side Management 1997," May 1999, pp. 1-2.

Internet: www.aceee.org/pubs/u951.htm, Geller, Howard and Steve Nadel and Miriam Pye, "Demand-side Management at a Crossroads," May 1999, p. 1.

Internet: www.buildingsmag.com/article 513.asp, Raiford, Regina, "Enlightened Environments," June 1999, pp. 1-3.

Chapter 3

Control Technology, Microelectronics and Nanotechnology

Control systems require an understanding of the characteristics of the process and involve the following:

- controlled variable,
- manipulated variable and
- uncontrolled load.

The control loop is closed by installing some type of automatic controller onto the process. The control loop must be able to measure the controlled variable and compare that measurement to some desired target value such as a temperature setpoint. If there is an error and the controlled variable (temperature) does not equal the temperature setpoint, then the output is modified and this changes the manipulated variable in such a way to reduce the deviation from setpoint.

The oldest control approach is to use a simple switch for control. This switch might turn on the heat source when the temperature is low and turn it off when the temperature is high. This type of system is called two-position control and is shown in Figure 3-1.

Controllers have different operating modes to meet the needs of the particular process. These modes are tuned to optimize the control function. Temperature control is a relatively slow, nonlinear (variable gain) process. The process is not noisy. Gas pressure control has no noise, and it is linear. A proportional-only controller can be used. When proportional-integral (PI) control is used, the integral is set for 1/3 to 1/5 of the period of oscillation which is usually in seconds.

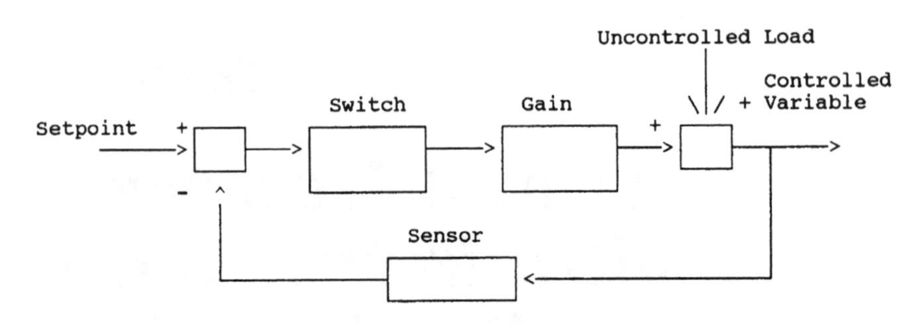

3-1. Two-position control system.

Loop Gain

Oscillations will result if the control loop gain is greater than 1.0. These oscillations will grow in amplitude until some component in the loop is destroyed. This condition is known as an unstable loop with runaway oscillations.

At a control loop gain of exactly 1.0, the loop will oscillate, but the oscillations will not grow in amplitude. The loop will be in a state of marginal stability.

Quarter amplitude damping results if the loop gain is 0.5. The loop response to an upset will be a damped sinusoid with the second peak at 1/4 the amplitude of the first peak. The goal of tuning in most process control loops is to obtain quarter amplitude damping. This will result if the product of all the gains in the loop comes to 0.5. It can be achieved by tuning, which adjusts the controller to make the overall gain product 0.5.

The controller gain, as well as most other gains, consists of a steady-state component, Kc which is the proportional setting. It is unaffected by the period of oscillation. The dynamic gain, Kd, varies with the period of the input (error) sinusoidal wave. The Kd × Kc product is the total gain of the controller.

Slow processes can be controlled with high-gain controllers, while controllers on fast processes must have low gains. An example of a slow (low-gain) process is space heating, where it takes a long time for the accumulation of the heat input provided by the manipulated variable to cause a small change in the controlled variable, the space temperature. If the combined gain of that process is 0.01, the controller gain required to provide quarter amplitude damping is 50 since $0.01 \times 50 = 0.5$.

If the controlled process is a flow process, the process gain is high, maybe 10.0. In this case the controller gain should be 0.05 ($10.0 \times 0.05 = 0.5$), so that the total loop gain is 0.5.

The higher the controller gain, the more corrective action the controller can apply to the process in response to a small deviation from setpoint, and the better the quality of the resulting control will be.

When the gain product of controller and process reaches unity, the process becomes unstable and undampened oscillations (cycling) occur. It is more difficult to tightly control fast (high-gain) processes without some oscillating. It is far easier to obtain tight control on slow, low-gain processes, since the use of high-gain controllers will not affect stability as much.

The loop gain is the product of all the gains in the loop, including sensor, controller, control valves and process. In a properly tuned loop, the product of all these gains is 0.5. What makes tuning difficult is that the process gain often varies with load.

In heat transfer processes, when the heat load is low and the heat transfer surface available to transfer the heat is large, the transfer of heat is performed efficiently. This type of process is a high-gain process. As the load rises, the heat transfer process becomes a low-gain process since the fixed heat transfer area becomes less and less sufficient to transfer the heat. This is why the gain of a heat transfer process varies with load.

Tuning such a system can be difficult since in order to arrive at an overall loop gain of 0.5, the controller needs to apply a high gain when the load is high and a low gain when the load is low. Standard controllers cannot do that, because they are tuned to provide a single gain.

If the loop is tuned at high loads, the loop can cycle when the load drops. If the loop is tuned at low loads, the loop will not be able to hold the process on setpoint when the load rises. It will be slow or sluggish to respond.

One way to compensate for the variable load effect is to use a variable gain control valve in the loop, which increases its gain as the load rises. If the gain of the process drops, the gain of the valve increases and the total loop gain remains relatively about the same.

Feedback Control

Automatic control systems can use either feedback (closed-loop) control or feed-forward (open-loop) control. Feedback control is more commonly used and is the underlying concept on which most automatic control theory is based. Feedback control is designed to achieve and maintain a desired condition by using the following control actions:

- measuring the process condition,
- comparing the measured condition with the desired condition, and
- initiating corrective action based on the difference between the desired and the actual condition.

The feedback technique is similar to the actions of a human operator attempting to control a process manually. In the control of a direct contact hot water heater, the operator would read a temperature indicator in the hot water tank and compare its value with the temperature desired. If the temperature were too high, the tank heat would be reduced, and if the temperature were too low, it would be increased.

An automatic feedback control system operates in much the same way. The temperature of the hot water is measured and a signal is fed back to a device that compares the measured temperature with the desired temperature. If an error exists, a signal is generated to change the heat value in such a way that the error is eliminated.

The main difference between the manual and automatic means of controlling the heater is that the automatic controller is more accurate and consistent. Both systems contain the basic elements of a feedback control loop.

Feedback control advantages over other techniques such as feed-forward control include its relative simplicity and potentially successful operation even when there are unknown contingencies. It works well as a regulator which maintains a desired operating point by compensating for various disturbances that affect the system. If designed and tuned properly, it works equally well as a control to initiate and follow changes demanded in the operating point.

Feed-forward Control

Feed-forward control is an alternate basic technique used to compensate for uncontrolled disturbances entering the system. In this technique the control action is based on the state of a disturbance input without reference to the actual system condition.

Feed-forward control should yield much faster correction than feedback control. In the ideal case compensation takes place in such a manner that the effect of the disturbance is never seen in the output.

A competent operator could use feed-forward to offset the changes in inlet water temperature of a direct contact water heater. Detecting a change in inlet water temperature, they would increase or decrease the heat rate to annul the change. The same compensation could be done automatically with an inlet temperature detector designed to initiate the proper corrective response in the heat value.

The notion of feed-forward control is forceful, but it can be hard to implement in a pure form in most control applications. In many cases disturbances cannot be accurately measured and this makes pure feed-forward difficult to use.

A major limitation of feed-forward is the inability to generate perfect models or to make accurate measurements. Because of these limitations, pure feed-forward would accumulate the errors in its model and would eventually self-destruct.

The main limitation of feedback control is that it cannot predict upsets but can only make corrections after the upsets have occurred. It makes this correction in an oscillating, cycling way.

The best use of feedback and feed-forward is to combine them in such a way that the imperfect feed-forward model corrects for about 90% of the upset as it occurs and the task of correcting for the remaining 10% is done with the bias generated by the feedback loop. Using this approach, the feed-forward component is not pushed beyond its abilities and the load on the feedback loop is reduced by an order of magnitude, allowing much tighter control.

Proportional Control Mode

Controllers generally use different control modes for control management. The proportional mode is the simplest of these. It is characterized by a continuous linear relationship between the controller input and output. Other terms in common usage include proportional action, correspondence control, droop control, and modulating control. A proportional control loop is shown in Figure 3-2.

The adjustable parameter of the proportional mode, Kc, is known as the proportional gain, or proportional sensitivity. It may be expressed

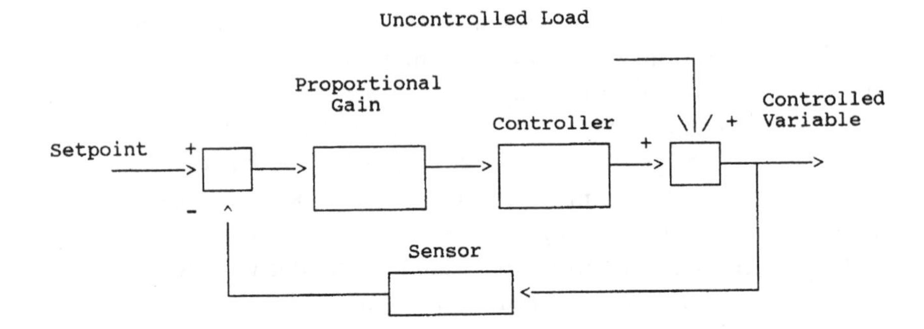

Figure 3-2. Proportional-only control.

in terms of percent proportional band (PB), where

Kc = 100PB

A wide band, high percentage of PB corresponds to less sensitive controller settings, and a narrow band, low percentage, corresponds to more sensitive controller settings.

The correction generated by the proportional control mode is proportional to the error. The proportional controller responds only to the present. It cannot consider past history of the error or possible future consequences of an error trend.

It responds to all errors in the same manner, in proportion to them. When a small error results in a large response, the gain (Kc) is said to be large or the proportional band is said to be narrow. When it takes a large error to cause a small response, the controller is said to have a small gain or a wide proportional setting.

The gain in DCS control units is usually adjustable from 0 to 8, while in analog controllers the range can usually be adjusted from 0.02 to about 25.

Proportional Offset

The main problem with proportional control is that it cannot keep the controlled variable on setpoint. The proportional controller can only respond to a change by moving away from its setpoint. The process will be on setpoint for only part of the time. At all other times the operating level travels up and down on its operating line as a function of the load.

The difference between the actual value and the setpoint is called offset. This is the amount by which the process is off setpoint.

The gain of the controller can be increased which narrows the throttling range of the system and makes the operating line more horizontal. Increasing the gain, can reduce, but not do away with the offset. At high gains the presence of the offset is no longer appreciable and it seems as if the controller is keeping the process on setpoint.

Most control loops become unstable if the controller is provided with a high gain. The exceptions are very slow processes. This limits the use of plain proportional control to processes which can tolerate high controller gains, the narrow proportional bands found in thermostats and humidostats. The addition of an internal feedback mechanism effectively reduces the sensitivity of the device to an acceptable range for proportional control.

Integral Mode

This mode is also known as reset mode, since after a load change it returns the controlled variable to setpoint and resets or eliminates the offset which the proportional controller would cause. This mode is also known as floating control. But, it is most often called integral control since the controller output is based on the integral of the controller input.

Proportional-plus Integral Control

In a proportional-plus integral controller the controller output is a combination of the integral and proportional value (Figure 3-3). The integral term is divided by the integral time setting of the controller, which is also called the reset time.

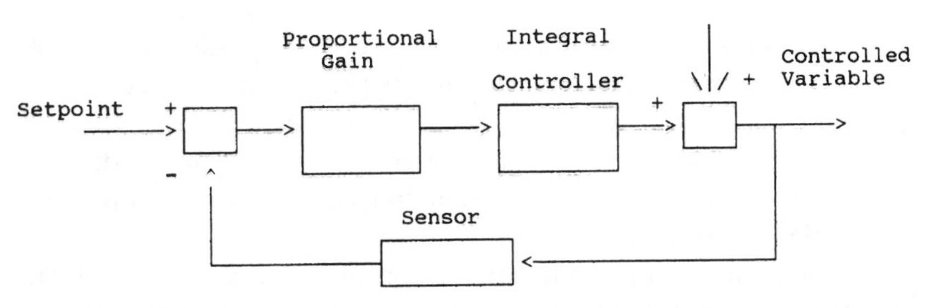

Figure 3-3. Proportional-plus-integral control.

The integral mode eliminates the offset which exists in proportional control. Proportional control results in an offset since it disregards the past history of error and leaves the accumulated effect of past errors uncorrected.

The integral mode continues to use the total past history of the error by integrating the area under the error curve. This integral correction acts to eliminate the offset.

The integral time is usually given in repeats/minute or in minutes/repeat. The integral setting of control loops in DCS systems is usually set from 0 to 300 repeats/minute, or from 0.2 seconds to about 60 minutes or more in terms of minutes/repeat. The shorter the integral time, the more often the proportional correction is repeated (more repeats/minute).

Pure integral control or floating control is rarely used. In very noisy valve position or flow control systems, the PI loop may be tuned with low gain and high reset (integral). The proportional mode tends to act as a noise amplifier, while the integral mode integrates the area under the noisy error curve and provides a smoothed average.

Reset Windup

While the integral mode solves the problem of offset, it introduces another problem. When the loop is idle, if the integral mode is allowed to stay active, it will eventually saturate, and its output will either drop to zero or rise to the maximum value of its power supply. Once saturated, the controller is not able to control and can upset the process by trying to introduce a correction for the error, which it experienced during its idle state.

In these installations the controller must be provided with either an external reset, which prevents it from becoming idle, or with anti-reset windup, which protects it from saturating in the idle state.

In some control systems, when a windup limit is reached, the integral time (repeats/minute) is increased to speed the unwinding and return the system to normal operation. In DCS systems these operations are implemented in software.

In digital systems the integral is approximated by an accumulator which continuously adds the deviation from setpoint to the previous sum. When an error exists over a long period of time, the integral can grow to the maximum value the DCS memory can hold. This condition is integral or reset windup.

The controller output in this case will be dominated by the wound-up value and will saturate with its control value either fully open or fully closed. Even when the error is removed or changes sign, the wound-up value continues to maintain its saturated output for a period of time. The saturated value will continue until the negative error accumulation equals the previously accumulated positive error.

When the controller is not in control, the integral will windup. This can be initiated by the following:

- failed control-loop communications,
- sensor failure,
- control constraint, and
- process off or not working properly.

Integral controllers should have an antiwind-up mechanism. One technique is to specify a limit on the integral contribution to the total output signal. Another strategy is to inhibit integral accumulation if the output reaches a limit. Either mechanism limits the integral windup to a value corresponding to full open or full closed. To accelerate the return of the integral contribution to its proper value, some controllers make the integral accumulation 10 to 20 times faster when coming off a limiting value.

Reset Feedback

Reset feedback is another technique where the integral of the error is replaced with a filter on the output. In reset feedback there is no integral to windup. But, there can still be some windup to the upper or lower limits of the output.

The filter can be on an external signal which would be identical to the controller output if the control loop were closed. There is a windup to the edge of the saturation value and when control is recovered there is no need for wind-down acceleration.

Derivative Mode

The proportional mode considers the present state of the process error and the integral mode uses its past history. The derivative mode anticipates its future state and acts on this forecast. This third control mode became needed as the size of equipment increased with a greater mass and thermal inertia.

A large process has more inertia or momentum which makes it difficult to stop or reverse a control trend. The derivative mode tries to predict process errors before they evolve and take corrective action in advance of the occurrence.

When the error is constant, the derivative contribution to the output signal is zero. This is because the derivative contribution is based on the rate at which the error is changing. When that rate is zero the derivative is zero. The derivative contribution always corresponds to the slope of the error curve.

Derivative Time

The derivative setting uses a derivative time (Td) which is the length of time the derivative mode looks into the future. If the derivative mode is set for Td, it will supply a correction that will equal the correction which the proportional mode would have generated at Td. The longer the Td setting, the further into the future the derivative-mode predicts and the larger is the corrective value. The derivative time can be adjusted from a few seconds to hours.

PD control loops are sometimes used in the slave controller in temperature cascade systems when there is a need to increase the sensitivity of the slave loop. PID control is more widely used and applications include many temperature control systems.

Limitations of Derivative Mode

The derivative mode acts upon the rate of error signal change and it can cause unnecessary upsets. It tends to react to sudden setpoint changes and will amplify noise. The control algorithm can be altered so that the derivative acts on the measurement and not on the error. This will reduce upsets.

Excessive noise and step changes in the measurement can be corrected by filtering out any change in the measurement that occurs faster than the maximum speed of response of the process. DCS systems provide software with adjustable filters for each variable. The time constant of these filters is usually adjustable from 0 to 100 seconds. In analog control systems, inverse derivative modules are often used.

Inverse Derivative Mode

This control action is used on fast processes. The inverse derivative mode is the opposite of the derivative mode. While the output of the derivative mode is directly proportional to the rate of change in error, the output of the inverse derivative mode is inversely proportional to the rate of change in error.

Inverse derivative is used to reduce the gain of the controller at high frequencies and is useful in stabilizing a control loop. The dynamic gain of the derivative function is selected as 0.5 or less. The gain of the inverse derivative controller decreases from the proportional gain at low frequency to a limiting value of the proportional gain divided by this factor at high frequency.

A proportional-plus-inverse-derivative controller provides high gain to minimize offset at low frequency and low gain to stabilize the loop at high frequency. Inverse derivative can also be added to a proportional-plus-integral controller to stabilize loops requiring very low proportional gain for stability.

Inverse derivative should only be added when the loop is unstable at the minimum gain setting of the proportional-plus-integral controller. It is available in a separate unit and can be added to the loop when stability problems occur. The addition of inverse derivative when properly tuned has little effect on the natural frequency of the loop.

Digital Control Modes

When control mode algorithms are implemented in digital systems, integration is replaced by finite summations and time differentials become finite differences. The scan period of DCS systems is usually fixed at 0.5 seconds or selectable for each loop at 0.1 to 30 seconds.

The DCS system does not continuously evaluate each measurement, but checks them intermittently. It makes its approximations on the basis of the present and the previous error. One type of calculation in use is called positional where the full output signal is recalculated every time the measurement is checked. The derivative is approximated by the change between the previous and the present value of the error. The integral is approximated by the sum of all previous errors between time zero and the present time.

Another type of calculation is called a velocity algorithm. When it is used, the value of the previous output signal is held in memory, and only the required change in that output signal is calculated. The change in the error between the previous and the present scan period is calculated (second derivative).

The positional algorithm is used with noisy measurements since it works with the error and not the rate of error change. Velocity algorithms can provide bumpless transfer, less reset windup and are better suited for motor-driven devices. Their main problems are noise sensitivity and oscillation.

Setpoint Response

When the error is the deviation from setpoint, a sudden setpoint change causes a sudden error change, which can cause a controller output spike. When both proportional and derivative control act only on measurements, the output spiking due to setpoint changes is prevented while the tuned controller performance is not affected. Some controllers have setpoint ramping to prevent output spiking by eliminating the sudden setpoint change.

Setpoint Stations

Setpoint stations are used to interface a digital control computer with analog controllers and actuators (Figure 3-4). They allow the computer to make changes in operating conditions. In supervisory control applications, the computer adjusts the setpoint of an analog controller. In direct digital or DCS control, the computer may adjust the actuator directly. The signals from the computer must be translated into the signals required by the receiving device.

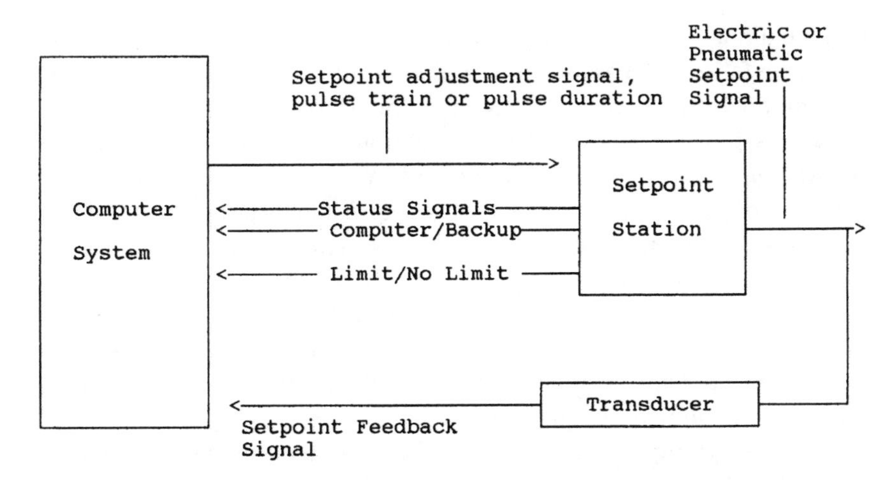

Figure 3-4. Setpoint interface signals.

Setpoint stations may be designed to interface with electronic or pneumatic controllers. The setpoint station may generate, from the computer output signals, a setpoint that can be directly used in the controller. For analog electronic controllers, a voltage output is generated by the computer (through a digital-to-analog converter) which is proportional to the desired setpoint.

Computer Setpoints

The operator can adjust the effective setpoint or use the computer output signal for computer control. Without automatic switching, this scheme is not fail-safe, since a computer system failure and loss of the computer-supplied setpoint voltage could cause a zero setpoint to exist. This could cause undesirable and even dangerous conditions to exist. There is a need to have automatic switching to a back-up setpoint adjustment in the event of computer failure.

The computer must be able to continually adjust the back-up setpoint to keep it up to date with the computer's so that a bumpless transfer is guaranteed. Then, when the transition from back-up control to computer control is made, the computer must supply a signal identical to that supplied by the back-up, in order to achieve a bumpless transfer. Most setpoint stations are designed to provide fail-safe operation in case of computer failure by having the computer supply a live signal only while making changes in the setpoint.

One fail-safe technique involves operating a stepping or synchronous motor with the computer digital output signals. The stepping motor receives a pulse train from the computer and the number of pulses is equivalent to the change in setpoint to be made. Failure of the computer results in no more pulses and leaves the setpoint in its last position.

A synchronous motor can accept a timed contact closure signal from the computer. Any change made in the setpoint is proportional to the duration of the contact closure signal. Unless the computer system fails while changing the setpoint fail-safe operation is possible. Adjustable mechanical stops can be used to keep the computer setpoint within limits and minimize the effects of such a failure.

In a stepping motor system, resolution is the number of steps required to drive the setpoint from zero to full scale. A full-scale travel of 100 steps, for example, gives a resolution of 1.0%. The maximum pulse rate is the sum of the minimum on pulse duration and the minimum allowable time between pulses. For example, if the sum is 30 milliseconds, the maximum pulse rate is 33 pulses/second, and, assuming 1.0% resolution, 3 seconds would be required for a full-scale change (100 pulses) in setpoint.

Another technique for generating consistent setpoint signals is to use an integrating amplifier. The output of the amplifier follows the setpoint and holds an equivalent setpoint reference voltage. The amplifier drift rate must be considered if the computer is not adjusting the setpoint often.

Setpoint Feedback

In a supervisory control system, the computer program must be initialized prior to taking over control of the setpoints. In either supervisory or direct digital control the computer needs to know the results of every setpoint adjustment so setpoint feedback signals are sent to the computer. These feedback signals can be obtained from a transducer.

Supervisory Control

In 1957, the Ramo-Woodridge Corporation introduced the first solid-state digital computer for control systems. This computer was designed around a magnetic drum with 8064 words of 20-bit length. The external input/output supported 540 1-bit digital inputs. The computer was designed to interface directly with existing analog control systems. The system acted as an enhancement to the existing control instrumentation and it functioned as a supervisory control device.

The software program consisted of models of the process. These were usually simple linear programming algorithms. The models provided new sets of operating conditions (setpoints) which could improve performance. The process was moved from one set of steady-state conditions to another in response to changes in uncontrolled disturbances such as ambient conditions.

Within a few years, other companies introduced their versions of control computers. These companies included IBM, General Electric, Control Data Corporation and Foxboro. The initial improvements involved increased memory size. The use of magnetic core memory increased the computational speed from milliseconds to microseconds. A major step came with the replacement of solid-state circuitry (diodes, resistors, and transistors) with integrated circuits. This technology provided increased speed and capacity, lower power requirements, smaller size and improved reliability.

An evolution in control computers came with the development of direct digital control (DDC). Here, the function of the analog instrumentation was incorporated in the computer and the analog controllers were eliminated. The goal was to reduce the control system cost and increase design flexibility for new applications.

Initially, these computers and especially the software, which had become complicated, were not sufficiently reliable. In many cases, when the computer went down, control was lost completely.

When minicomputers replaced these older computers, it was feasible to provide redundancy in the form of dual computers. This solved the reliability problem.

DCS

In 1976 when Honeywell announced the first distributed digital control system (DCS), the system was based on redundancy using microprocessor-based controllers, redundancy in communications and redundancy in operator interfaces.

Most large-scale control applications now use microcomputer-based DCS (Figure 3-5). These systems generally have CRT-based operator consoles and keyboards which are used for monitoring and controlling the system. Controllers and multifunction control modules are used to provide the basic control computation or operation.

A communication network transfers the information between the control modules and operator consoles. I/O (Input/Output) modules are used to convert the sensor signals from analog to digital and digital to analog for the controller modules and console displays. Fieldbus communication links are used for communication between remote I/O devices and control modules. Historical data storage is used for control data and is available for on-line data retrieval. A control software package is used for monitoring, control, reporting, graphics and trending.

In most applications, a DCS replaces a conventional control system for the purpose of performing basic control. The supervisory control functions performed by the older computer systems plus additional functions such as on-line information evaluation may be performed by a supervisory computer linked to a DCS.

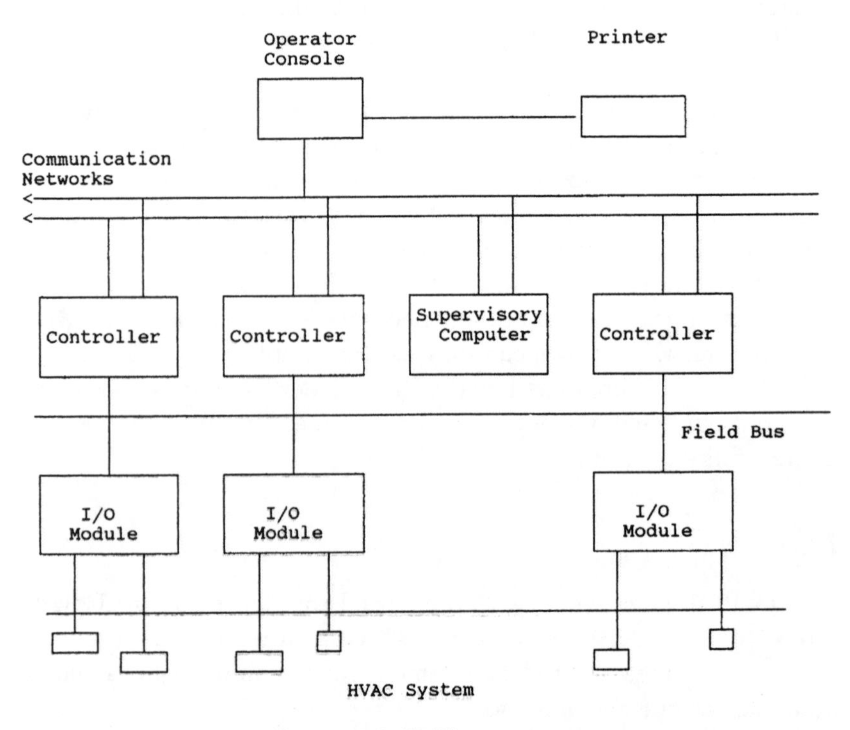

Figure 3-5. DCS structure.

Supervisory Functions

The use of digital computers and their capability to perform complex mathematical calculations and make logical decisions provide the opportunity to improve the performance of any control application. The need for supervisory control results from the fact that all processes are affected by changes. These changes are in the form of disturbances. These disturbances may come from many sources. Examples of disturbances include ambient conditions, heater and heat exchanger fouling and economics. The supervisory control function can also be used off-line as a valuable simulation tool for evaluating what-if scenarios.

Alarms are based on models utilizing the present and past history of the data and the control commands. Anticipating alarm conditions in advance is a vital function of supervisory control. The supervisory computer typically determines the operating constraints, such as flooding or surge conditions. Basic material balance, energy balance, or heat transfer calculations are utilized for model creation.

The computation of the model is not necessarily a single-path computation. It may require iterative computations. This type of computation is not usually performed at the DCS level.

The present operating state is determined based on the on-line, real-time information from the DCS using temperatures, pressures and other characteristics. This determines the desired state according to the constraints and optimization criteria. An on-line information system may include data collection, data reconciliation, data storage and retrieval. The optimum control strategy is based on the on-line, real-time information. The control strategy commands will follow the optimum path such that the objectives of minimum cost and minimum energy are obtained.

Control and Optimization

Supervisory control can include calculation of the present state of the process in terms of efficiency and calculation of new conditions, in terms of setpoints which will meet the requirements of the objective function. The objective function can maximize efficiency through fuel costs.

The tools needed to perform supervisory control are a model and an optimization or maximization algorithm. The model can range from simple linear equations to more sophisticated theoretical expressions. The main purpose of the model is to provide an estimate of the relativity between the manipulated variable and the intermediate and performance variables. When the manipulated variables are perturbed by the maximization algorithm, an estimate of the change in the intermediate and performance variables with relation to the constraint values must be made.

The optimization or maximization algorithm can also be simplistic or sophisticated. There is an advantage to using the simpler approach. The procedure always starts with a known solution. Then, the perturbation procedure takes advantage of what a digital computer does best, simple calculations repeated over and over. In sophisticated optimization algorithms, there may be multiple solutions and the real solutions must be selected.

In general, the computers used for supervisory control are multitasking computers which are interfaced with the DCS through a computer gateway (Figure 3-6). The data which reside in control mod-

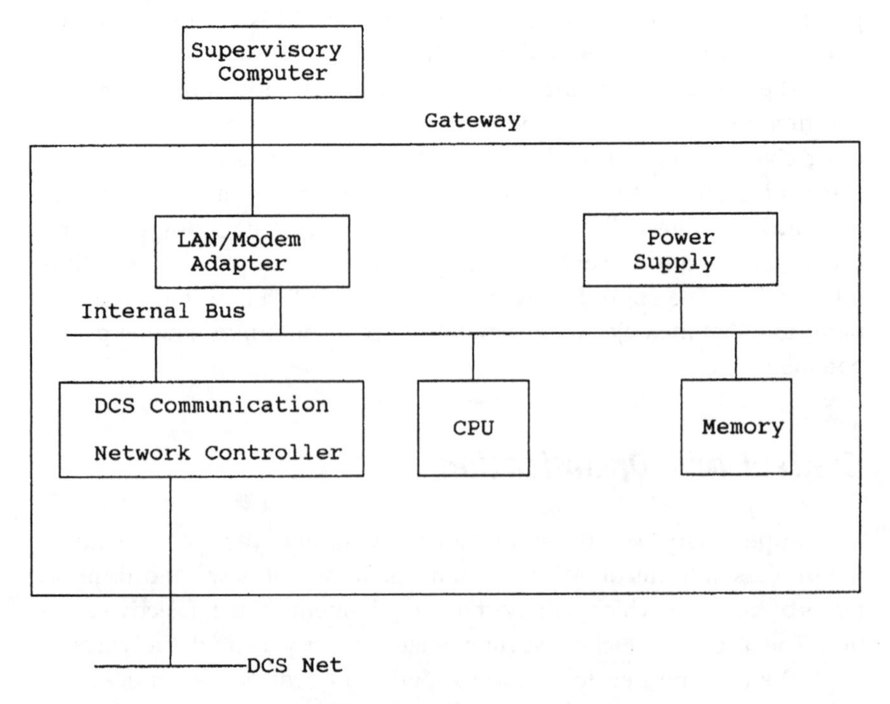

Figure 3-6. Gateway for DCS.

ules can be communicated to the supervisory computer when this information is needed for computation.

Supervisory Techniques

Early supervisory control was used to command single-loop analog controllers in order to achieve a certain goal, such as to obtain a uniform temperature or to determine the optimum blending of cold and hot decks. The functions that the analog controllers could not accomplish were delegated to the digital computer to perform. These included the logic, sequence, or analytic computations to improve the control operation.

Coordinated supervisory control with integrated feedback or feedforward combined in a velocity control algorithm may be activated or deactivated without creating perturbations. The velocity algorithm implemented in the supervisory computer can be converted to the position algorithm which is usually applied at the DCS level. The control command, in the supervisory computer at the current sampling time is made up from the current computer position command and the past sampled position command.

Anti-windup Control

The command from the supervisory computer to the setpoint for the control loop at the DCS level is a form of cascade control. The primary loop control command (supervisory computer command) requires information from the lower or secondary loop at the DCS level in order to command correctly. This information includes saturation status and lower loop current setpoint values at the DCS. This information at the DCS level requires constant monitoring by the supervisory computer of the real-time information on the DCS network. It is important to prevent the computer command from causing setpoint windup at the DCS level.

Protecting the upper loop command from windup involves a freezing of the supervisory control command output to the DCS setpoint at its last value if the DCS output to the valve is saturated and the direction of the upper loop output increment is such that it would cause further valve saturation at the DCS level. Otherwise, the upper loop in the supervisory computer will not be frozen at its last value.

For supervisory control, the upper-level control algorithm must be initialized when the output from the supervisory computer commanding the lower loops in the DCS is initiated or when an emergency condition requires that the supervisory control be prevented.

System Failure

When the supervisory computer fails, the control generated at the computer level cannot be transferred to the DCS level. An integrated control system should use a watchdog timer or communication link message to flag the lower DCS loops. This is needed to prevent an unnecessary disturbance to the system due to a supervisory computer failure.

Fault-tolerant control can be achieved in the event that the computer fails. One approach to fault tolerance is to implement a generalized output control algorithm at the DCS level.

Supervisory Displays

The display on the CRT screen becomes the main interface which can be used for system performance evaluation. There are usually hundreds of display page screens provided in both DCS and supervisory computer software. These display pages are easily accessed through a keyboard, touch screen, mouse, or track ball.

A typical display access method is to use a tree-structured, multilevel display architecture. A two-level structure with 16 elements (pages) in each level provides 16 × 16 = 256 pages. Each screen page at the second level can cover 16 elements and each element can be a digital or analog input/output combination. This provides a total of 4096 combined I/O points that can be displayed at the second level. If the proper search path is selected, only two keystrokes are needed to arrive at the desired point.

Traditionally analog controllers have been arranged in clusters of rows or columns. In DCS, controllers are usually arranged in groups of 8, 12, 16, or 32 for screen displays. One of the simplest ways to represent information for supervisory controls is to use a multi-input/multi-output display.

Alarms

Alarms for DCS include high, low, high-high, low-low and high rate of change. These alarms appear on the alarm screen at the DCS operator console along with time of occurrence, magnitude of variable and alarm limit. In supervisory computers, an alarm management program can assist the alarm diagnostics and help the operator pinpoint the alarm cause and location. The alarms can be grouped according to the geographical regions of the plant.

Voice input and output can be important for conveying information about real-time conditions. The cause or location of alarms can be announced without using the screen display. When variables approach an alarm limit, voice output can be generated to alert an impending alarm condition.

The periodic execution of the program is typically a function of dynamic parameters such as time constants, transportation delays, program execution time, network speed, gateway update frequency and periodic interrupts.

Gateway Interface

Supervisory computers are interfaced to the DCS through a gateway device which receives data and messages from an originating source. The gateway resides on a node of the local control network or data communication network of the DCS (Figure 3-7).

The gateways are configured through the DCS using a gateway data table. Data in the data table could include the following types:

- setpoints,
- process variables,
- control loop outputs,
- control parameters, and
- alarm status.

The gateway device collects the listed data from the DCS communication network or data highway and transfers them to the supervisory computer through a different communication link or network. The gateway receives data from the supervisory computer and transfers these to

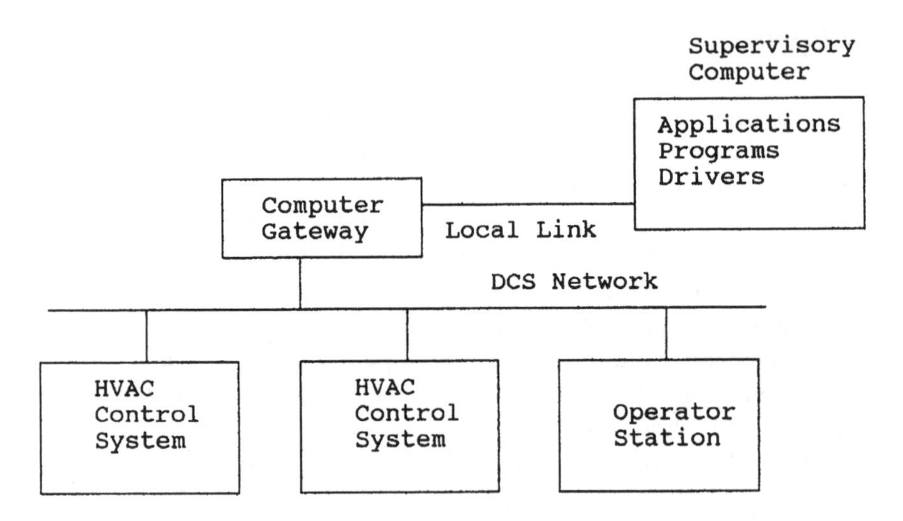

Figure 3-7. Supervisory computer and DCS gateway.

the nodes of the DCS. Gateways are also used to communicate with computers through a local area network (LAN).

Old and New Control Concepts

The software used in modern building control systems can result in significant benefits. More control in these facilities can minimize expenses. However, the sophisticated building control systems require some changes. One change is to avoid implementing direct digital controllers to deal with the same inputs that were used in past generations of pneumatic controls. A more holistic approach is required in modern building controls.

Today's building control systems are fueled by software. If properly applied, this software can deliver powerful results. But, this requires breaking down some traditional lines of system separation.

Direct digital control can transform HVAC controls. Just a few years ago, a building's HVAC control system might consist of a complex network of discrete electro-mechanical components which were linked together with pneumatic tubing. Each component had a specific function, either serving as a link to the physical world, or being part of the control logic itself.

Component Control

In the logic components, decisions were based on the physical parameters connected to it, such as air pressure in a pneumatic control line coming from a sensor. Every sensor required pneumatic tubing to be run back to the controller. If a change in control logic was desired, it required additional labor and materials. This type of control focused on optimizing components, instead of looking at the dynamic ways of optimizing the control operation.

In a DDC building management system, any point connected to the system is a part of the data network, capable of providing critical control information.

A component concentration prevents DDC control from using information from anywhere on the network. This can occur if it is implemented only to consider the same inputs that were used by the systems of pneumatic controls.

Temperature control of a single office in a large complex can illustrate this point. The office temperature might be controlled by a DDC variable air volume (VAV) box controller with a sensor/thermostat on the wall. To save energy, a night setback feature is added when there is no one in the office, with the change-over triggered automatically by a time scheduler.

The individual office has an installed light fixture and wall switch. To save energy, an occupancy sensor is mounted in the room that turns off the light when no one is there.

This results in two individually optimized systems that satisfy their own individual missions of saving energy when the room is not occupied. But, has the office, as a whole, been truly optimized?

In an integrated building management system, every piece of datum is available for use in any strategy and an occupancy sensor to control the DDC VAV box should be used. The occupied/unoccupied schedule and temperature setpoints are still used, but during occupied hours, a third standby setpoint is used for damper control of the VAV box. This would allow the temperature to drift up or down to the predefined standby temperature setpoint. This setpoint is used during normal occupied hours when the occupant is out of the room. The result is more energy savings at little additional cost.

This occupancy sensor could be used to provide security information as well. Using data generated by the system, security could tell which rooms were occupied at any given time.

Microelectronics

The most important single invention within the whole complex of inventions which is today called computer technology is the transistor. Electronic tubes made the first high-speed processing possible, and the stored program led the way to the possibility of computer intelligence, but the transistor paved the way for modern computers.

Tubes rely for their amplifying and switching power on a heater electrode which pumps electrons through a vacuum. This electrode must be shaped out of metal and it cannot be reduced beyond a certain size. If it gets too small, it cannot produce enough heat to activate the electrons.

The transistor relies for particular structures of a minute size which are formed inside silicon crystals. These structures can be very small and a solid-state amplifier or switch may be based in a tiny fragment of silicon. This results in substantial reductions in the size of the functioning unit of an amplifier or computing unit.

The first transistorized devices occupied less than a hundredth of the space of electronic tubes. Because they do not rely on heat to drive their electrons along, transistor devices need no warming up period and they consume far less energy. They are also faster in operation and much more reliable. Transistors quickly replaced electronic tubes except for some high-power applications.

The coming of the transistor could not have been foreseen by any of the computer pioneers, nor could its dramatic consequences have been anticipated. The power and speed of all-transistor computers rose to a point where they were quantitatively different from anything that the early computer pioneers could have contemplated.

Computer memories became larger, moving from a few hundred or thousands of bits of the earliest devices to hundreds of thousands, millions, tens of millions and eventually billions of words. These massive memories, now available for low cost personal computers, began to take on new and unexpected roles. Instead of just being heavy-duty calculators, they suddenly became information handlers as well offering huge and progressively cheaper repositories for masses of information. As memory space continued to multiply, memory banks became lighter and smaller, their processing speeds increased and costs continued to drop.

The first components of the earliest computers were large and mechanical. Then came electro-magnetic relays, which were a little smaller, after which came tubes which were somewhat larger.

The arrival of the transistor produced a sensational reduction in size. The transistor itself is a spot of semi-conducting material. This material is not as good a conductor as metal, but it is better than an insulator like plastic or rubber.

When it contains certain impurities in its structure, this semiconductor can act as an amplifier and switching device. These impurities can be very small.

The first transistors that were manufactured had an initial size of about one cubic centimeter. Several levels of miniaturization were quickly achieved.

Transistors were then combined into logic units. These electronic circuits consists of 20 to 100 components arranged as sets of logic gates for AND, OR and NOT functions. They were built on a chip of silicon about a centimeter square. These chips have grown to where they are now the heart of just about every electronic device including watches and the largest computers.

Miniaturization did not stop when it came to etching complete circuits on a chip. Using techniques known as large-scale integration (LSI), hundreds, thousands and even millions of individual units were amassed on one chip of semi-conductor material.

Still the process of miniaturization continues into the future. The building units of computers are getting smaller and smaller, shrinking into the infinities of the molecular world.

Computer Applications

The outcome of these reductions in cost and size, and the huge increases in memory capacity and reliability, allow computing circuits to be widely used in areas which would have been inconceivable before the advent of microprocessors. Many areas of special-purpose computing which have been available at high cost and with only marginal reliability in the past have also become cheaply available.

Computers have the capacity to read and store text, graphics and even human retina signatures and fingerprints. They can also recognize the human voice and understand spoken words and phrases. They also have the ability to communicate with the user by drawing diagrams or pictures on CRT screens and also by the use of synthetic speech and music. The music, words and phrases are created by the computer.

Computers also have the ability to recognize shapes. These developments are only the first strivings of computers to interact in a dynamic, rather than a passive way with their environment. Once a computer has the power to recognize a shape or pattern, it can be utilized in an emergency or fire control system which allows it to do something to prevent injury or harm to humans and facilities.

Image recognition can take place with a robot scanning camera which could recognize intruders or emergency situations and control a motorized device to rectify the problem. It might also be programmed to understand a few useful words such as start, go and stop spoken by any human voice.

Computer Evolution

The computer revolution rides on the cost-tails of successful products. There is always another generation of devices capable of tackling a far wider range of tasks. The number-only keyboards and number-only displays of early calculators were replaced by full alphabetical hand-held personal computer devices.

The main difficulty of minuscule keyboards will be overcome when speech recognition systems shrink down to the point where they can be built into wristwatches and other small devices. Buttons as input mechanisms then become largely redundant. Calculations will be performed and telephone numbers and other data are entered into the memory by spoken instructions.

Computer voice output is also in an extremely rapid phase of development and speech synthesizers are diminishing in size and cost. One of the first applications was a talking calculator, which not only displayed the results of the calculation on its display but also offered them aloud in a synthesized voice.

Instruments may speak their measurements and containers may tell you how much they have in them. Talking altimeters and airspeed indicators may be used in aviation. In the home there will be speaking bathroom scales, freezers that remind you to restock them, ovens that tell you how the roast is doing, doorbells that specify how many visitors you had and when they came and thermometers which advise you what to wear before you get up. You can obtain electronic locks which open when you key in the appropriate combination and others that hold the combination on a plastic card.

In the automobile industry, cars use computer-controlled fuel injection which is extremely efficient and speedometers can compute your average speed and remaining fuel and tell you in synthetic speech when gas is low. Headlights can come on automatically when the ambient light falls below a certain level. Another use of computing is to calculate the speed of the car traveling in front and specify if the two vehicles are being driven at a safe distance.

Shrinking Devices

It still remains a question how small actual devices can be made. Existing systems can be extrapolated to the 50 to 100 nanometer region. As you start getting wires, transistors and other components closer together, they start to talk to one another, and this crosstalk becomes a serious problem. The properties of the basic materials used (doped silicon) also become hard to control.

It has seen from microelectronics that if you can make things smaller, there will be many applications for them. There is a need to make things small so they can be fast and cheap and portable and not power consuming.

Photolithography that uses ultraviolet light to etch out patterns on silicon chips is the dominant technology in making microelectronics on the micrometer scale. Photolithography is getting more complex and costly. The fabrication facilities for making semiconductor chips that are being planned now for the years past 2000, are estimated to be $3 billion to $10 billion. This means, that for a reasonable return on investment, a greater number of devices have to be sold every year for the few years that this facility is the state of the art.

Photolithography has been successful and there is every reason to think this will continue for a while. It is hard to go below 100 nanometers, and you cannot build 3-D things yet. It does not work for many materials other than silicon. One of the alternatives is electron beams. An example of this is Scalpel which is a system developed at Lucent Technologies' Bell Labs that uses electron beams to pattern silicon wafers.

Another alternative is X-ray lithography. This process uses X-rays to pattern wafers. Both of these have technical problems which will be worked out, but they may not be cheap enough in the end. Newer technologies, such as lithography, use neutral atoms or ion beams.

Nanotechnology

Shrinking microelectronics smaller and smaller into the nanometer region is called nanotechnology. This technology can be used for mechanical types of applications. One application could be very small machines that move around in the bloodstream. This would be similar to the flagellar motor in bacteria. One of the problems in this type of machine is power. Another example is sensors. Human sensors like the retina and the nose use molecules that are nanoscale sensors.

In the future, nanotechnology may change manufacturing as well as other areas. Micron-scale technology may cost less than newsprint. Instead of a newspaper, you might buy a sheet of paper. One side of it is a battery and the other side is a display. You can read it, scroll to find references, see animated illustrations and when you are done, throw it away. In about 10 years micron-scale electronics with new technologies may be used in toys, grocery labels, shipping labels, and credit/ID cards.

Modern technology heats up silicon, saws it into slices and makes patterns on its surface using stencils and light. This produces the electronic chips of microelectronic technology. It has moved machines as powerful as the room-sized computers of the early 1950s into pocket-sized computers. The individual components are 1/10th the width of a fine hair.

The new technology will handle individual atoms and molecules with precision and is sometimes called molecular nanotechnology. The terms nanotechnology and molecular technology are also used interchangeably to describe this new type of technology.

Microcircuits have parts measured in micrometers (millionths of a meter) but molecules are measured in nanometers, which are a thousand times smaller. The new technology will build both nanocircuits and nanomachines.

Molecular Nanotechnology

Early in the 21st century, we may have submicroscopic, self-replicating robots. Molecular nanotechnology could let us manufacture anything with atomic precision. Ten nanometers is 1,000 times smaller than the diameter of a strand of human hair.

In a computer, data are organized into combinations of bits, so they can be easily reproduced and distributed. With matter, the building blocks are atoms and the combinations of atoms that make up molecules.

Nanotechnology allows you to manipulate those atoms and molecules, making it possible to manufacture, replicate, and distribute any substance as easily and cheaply as you can replicate data on a computer. Any kind of material, from metal and wood to food and DNA could be replicated. This would be done with an assembler, a few-atoms-large nanomachine that could custom build matter.

The key to manufacturing with assemblers is self replication. If these assemblers could replicate themselves, there could be thousands or even millions of assemblers manufacturing in unison and manufacturing processes would be transformed.

Cornell, Stanford and Zyvex (the first molecular nanotechnology development company) have been working on such assemblers. Estimates of completion vary from 5 to 15 years. The first products could be superstrong nanoscale building materials, such as the Buckytubes envisioned at Rice University. Buckytubes are made from geodesic dome-shaped carbon molecules called Buckyballs, after Buckminster Fuller, the inventor of the geodesic dome. These tubes are basically nanometer-sized graphite fibers, with a strength of 100 to 150 times that of steel at less than 1/4th the weight.

Molecular Machines

Molecular machines will use molecular bonds that will be broken and reformed. Copper holds its shape because its atoms stick together in regular patterns. It can be bent and hammered because its atoms can slip over one another while remaining bound together. Glass shatters when it is hammered because its atoms separate before they slip. More complex patterns make up the active nanomachines of living cells.

Biochemists already work with these machines, which are chiefly made of protein, the main engineering material of living cells. Protein machines are unusually flexible.

In modern gene synthesis machines, genetic engineers build orderly polymers (DNA molecules) by combining molecules in a particular order. These molecules are nucleotides or compounds of DNA. The machine adds nucleotides in a particular sequence. It bonds one type of nucleotide to the chain ends, then washes away the leftover material and adds chemicals to prepare the chain ends to bond the next nucleotide. The chains bond on the nucleotides in a programmed sequence.

Protein Machines

Molecular machines are found in bacteria. They are also known as protein machines. Some function as restriction enzymes. These read the DNA sequences, by sticking to them and cut the chain by rearranging atoms. Other enzymes splice chains together. The enzymes are used together to write and edit DNA patterns.

These DNA patterns can be used to build proteins. The protein molecules can fold up to form small objects. Proteins range from human growth hormones to enzymes used in making cheese.

Programmable protein machines will take a large molecule (the workpiece) and bring a smaller molecule up against it in the right place. It will then bond the molecules together. By bonding molecule after molecule to the workpiece, the machine will assemble a larger and larger structure while controlling how its atoms are arranged. These protein machines will build small, solid objects of metal, ceramic, or diamond which are invisibly small, but extremely strong.

Some proteins can form basic mechanical structures like struts and parts of other molecules make excellent bearings. Amoebas and human cells move and change shape by using fibers and rods that act as molecular muscles and bones. Protein molecules will be used as motors, bearings, and moving parts to build robot arms which will themselves be able to handle individual molecules. Tiny motors and pumps may be made that would fit through the finest capillaries.

Molecular Computers

The Naval Research Laboratory, IBM and Genex Corporation have conducted experiments for developing molecular switches, memory devices and other structures that could be used in a protein-based computer. Japan has been developing self-assembling molecular motors and computers.

Bio-chips are another term for molecular electronic systems. NEC, Hitachi, Toshiba, Matsushite, Fujitsu, Sanyo-Denki and Sharp have started research efforts on bio-chips for bio-computers.

Molecular computers will control molecular assemblers, providing the flow of instructions needed to place large numbers of atoms. Nanocomputers with molecular memory devices can also store data generated by a process that is the opposite of assembly.

Assemblers will be used to shrink the size and cost of computer circuits. In today's bulk technology, the patterns on silicon chips remain flat and molecular-scale flaws are unavoidable. Assemblers allow circuits to be built in three dimensions with atomic precision.

The fastest computers will use electronic effects, but the smallest may not. With components a few atoms wide, a simple mechanical computer could be 1/100 of a cubic micron. This is billions of times more compact than today's microelectronics. Even with a billion bytes of storage, a nanomechanical computer would fit in a cube a micron wide, which is about the size of a bacterium or single bacteria. It would be fast. Although the mechanical operations will be about 100,000 times slower than electrical signals, they need to travel only 1/1,000,000 as far and there will be less delay. This results in a mechanical computer that works faster than most of today's electronic units.

Universal Assemblers

Second-generation nanomachines will use enzymes. These nanomachines will function as assemblers. Enzymes assemble large molecules by grabbing small molecules from the water around them and hold them together so that a bond forms. Enzymes can assemble essentially the whole range of molecules found in living things.

An enzyme-like machine could add carbon atoms to a small area, layer on layer. If bonded correctly, the atoms will form a fine, flexible diamond fiber having more than fifty times as much strength as the same weight of aluminum.

DNA copying machines make less than one error in 100,000,000,000 operations. This is done with DNA enzymes that proofread the copy and correct errors. Assemblers could use similar error-checking and error-correcting schemes. The future will bring many improvements on biological evolution, and nanomachines will be but one.

Advanced assemblers will be able to do more than existing protein machines. They will be programmable like ribosomes, but they will be able to use a wider range of tools. They will be made of materials that are stronger and more stable than proteins. They will be able to rotate and move molecules in three dimensions under programmed control, making possible the precise assembly of complex objects.

Disassemblers

Molecular tools will be able to take anything apart, a few atoms at a time. A nanomachine could also apply mechanical force, prying groups of atoms free. A nanomachine that is able to do this, while recording what it removes layer by layer, is a disassembler.

Assemblers can be used to synthesize things while disassemblers can help to analyze things. Assemblers require the ability of enzymes and chemical reactions to form bonds and of machines to control the process.

Disassemblers require the ability of enzymes and chemical reactions to break bonds and of machines to control the process. Enzymes, acids, oxidizers, alkali metals, ions, and reactive groups of atoms called free radicals can break bonds and remove groups of atoms.

Future Nanotechnology

As nanotechnology moves beyond proteins, molecules will be assembled like components. Molecular tools will bond molecules together to make tiny gears, motors, levers, and casings, and assemble them to make complex machines.

One type of virus acts like a spring-loaded syringe and can inject viral DNA. This DNA then directs the cell's machines to build more viral DNA and syringes.

Parts containing only a few atoms will be lumpy, but they can work if they have smooth bearings to support them. Some bonds between atoms make excellent bearings. A part can be mounted by means of a single chemical bond that will let it turn freely and smoothly. A bearing can be made using only two atoms and moving parts need have only a few atoms.

Assemblers, disassemblers, and nanocomputers will work together. A nanomachine will be able to direct the disassembly of an object, record its structure, and then direct the assembly of perfect copies. Molecular machines would serve a range of basic functions and the parts serving these basic functions may be combined to build complex machines.

In understanding the future, we need to understand the consequences of assemblers, disassemblers, and nanocomputers. They promise to bring changes as profound as the industrial and computer revolutions. Nanotechnology developments may sweep the world within 10 to 50 years.

Assemblers may take years to emerge, but their emergence appears to be almost inevitable. Although the path to assemblers has many steps, each step will bring the next within reach and each will bring its rewards. The first steps have already been taken, under the names of genetic engineering and biotechnology. Other paths to assemblers are possible. Advances in computer-aided design will speed the development of molecular tools and the advance toward assemblers will quicken.

These assemblers will be used in factory assembly lines. They will tailor many products such as clothes to personal specifications. Materials, from steel to cloth, will be built atom by atom, under the control of nanocomputers.

Huge leaps in medicine will enable many people to live to 120, or older. A scan of your genetic structure (with billions of bits of data in the estimated 80,000 genes in your body) will detect symptoms or susceptibility to particular diseases. Replacing diseased or worn-out body parts will be as routine as replacing auto parts today. Most diseases will be cured.

Nanobots or minuscule robots will deliver medication to affected cells to prevent or treat disease. They will clear clogged arteries and repair damaged tissue.

You will have your vital signs tested by machines at the drugstore and send the results to your doctor via the Internet for analysis. Implanted biochips may also monitor your signs, alerting you or your doctor to impending problems.

A remote surgeon will do hip and other replacements at your home. The doctor will view the surgical area on a screen and remotely manipulate surgical instruments inserted by a robot.

Neurons and Computers

When the first big computer attracted the attention of the popular press in the early 1950s, they were given the name of electronic brains. The human brain is made up of minute switching units called neurons. There are about 10 billion of these.

In the early 1960s, with transistorization, 10 billion electronic switches had shrunk to about the size of the Statue of Liberty, and a 10-kilowatt generator was needed for power. By the mid-1960s, integrated circuits compressed this electronic brain to the size of a Greyhound bus, and you could run it off a 220-volt outlet. By the mid-1970s it was the size of a TV set and the shrinking continues so now it is a little bigger than a credit card with most of the space taken up by the connections.

In the future, the nerve cells of the human brain may be linked to computer circuitry, creating a network that can interact with similar networks, and with computers and other databases. The contents of the brain could be scanned and downloaded into an external database, where they could be manipulated, stolen or even erased.

Future Technology

Other possible developments may affect control technology. At the Context-Aware Laboratory at MIT, nontraditional interfaces such as speech and gesture recognition are being combined with sensor data and specialized software models to allow computers to determine what humans are doing and act accordingly.

In the future, devices will be prevalent but unseen. They will be built-in or piped into homes, as utilities are now. These labor-saving devices will need no programming and batteries. Most household equipment will respond to voice commands. In an emergency you would give the command to get an ambulance and it would be done.

Smart cupboards and refrigerators will automatically reorder food that runs out. Clothing and household linens will be made of smart fabrics that clean and press themselves. Human washing machines will cycle you through soaping, washing, rinsing and drying.

Robots may pick up after us and eventually housekeeping will go beyond robotics and computers towards invisibility.

Personal Robots

There are already robots that will vacuum your carpet with the sweep of a mouse and carry your dishes to the dining room with the clap of your hands. The Cye Personal Robot is controlled by your PC to vacuum floors. Cye communicates with your PC using an FCC approved 900-MHz radio link. You map your environment on the screen so Cye knows the clear space and the obstacles.

This robot is controlled graphically from a PC. The window has a row of task types at the top, and you click and drag down into a flowchart. Tasks include

GO TO
WAIT UNTIL
SING NOTE
REPEAT
CHECK POSITION
VACUUM PATCH

The robot's radio pod is plugged into a COM port on your PC.

In the MIT Media Laboratory, there is a talking trivet, or hot pad, that senses the temperature of dishes it holds and announces if they are ready or need rewarming.

There is also a ceiling that has been turned into a display. This display could be used to check plant and building conditions. Doorways will detect visitors as they step onto a sensor-fitted mat and set appointments.

All-purpose flat screens will diagnose the failure of in-house systems and dial up real time images in all parts of the world for meetings and visits. Intelligent rooms will have walls that can see you with vision sensors, hear you by voice recognition systems and speak to you in response to your requests for heating and cooling. Hologram meetings, where projected images of people in different locations are used, may also become common.

Homes may evolve so they are completely self contained and mobile, so they can be moved anywhere. Electronic wallpaper will let you change the color or pattern of your walls on demand. Electronic image spots will display art that you dial up. Keys, locks and bolts will be replaced by voice commands to open doors and windows.

Vehicles of the future will be made of molded plastic and powered by non-polluting fuels. Other changes for motor vehicles include magnetized tracks for the interstates, which allow travel at bumper to bumper speeds of 200 m.p.h., with no real driving involved. Dashboard computers will warn of accidents and delays and tell the location of the nearest open parking spot. On-board computers will monitor the workings of your auto and diagnose incipient or actual failures, automatically informing the shop of spare parts you will need.

Aircraft will take off vertically, reducing noise and the size of airports. Orbiting the earth will be the first space cruise open to tourists. You will rise from ground to satellite on a space elevator, soaring up a tethered cable that generates its own energy. Orbiting hotels will quickly follow.

A two-planet interplanetary Internet may be in operation with several satellites in orbit around Mars. There may be as many as 2 billion users of the Internet and that many devices on the network as well.

Virtual reality and artificial intelligence will provide almost limitless entertainment offerings. Seated in a special chair that senses if you are cold or uncomfortable and adapts itself, you will pull a bubble screen around you. These computer/entertainment centers will provide virtual-reality simulations for training and entertainment.

Connectivity Trends

A connectivity revolution is taking place of once independent systems that are now connected. These evolving, interdependent systems provide solutions that save money while solving problems. An evolution of data delivery systems is taking place as offered by Internet or telephone suppliers.

In the new world of connectivity, the technology must not only deliver the data, but also get them to the right place, at the right time and at the right price. The information superhighway and the connectivity revolution are linked and have been evolving for about 15 years. Product life cycles were more simple then. A system might be designed in 2 years and manufactured for 5. Proprietary designs allow system modifications to remain in the with the manufacturer. These self-contained systems were closed boxes. They are now forced to communicate with outside systems. These systems must also accommodate a variety of connectivity standards that are evolving:

- IEEE 1394 for consumer electronics,
- IRBus for remote controls,
- IEEE 1451 for smart sensors and
- Universal Serial Bus (USB) or Device Bay for personal computers.

Connectivity is the key to the convergence of computer and communications electronics. It is also the key enabling technology in safety or security applications found in buildings. The connectivity revolution brings vast new options and new opportunities including a personal computer controlling HVAC equipment or an Internet remote monitor. Connectivity will change forever the way we view system applications. Open system connectivity, user flexibility and self-configuration in a more complex environment are becoming the path to the future. This means Internet access for remote control of lights and heat that allow the temperature to be right when you get to your office or home.

Historical systems for damper control might include a remote motor controller, motor and mechanical lever arm. Today's design might include air flow or pressure sensors and perhaps current sensing in the motor.

Security systems now typically use encryption and rolling codes for authentication. The security system may even sound an alarm if hackers try random codes to break in. There are many different technologies and possible ways to wire the system.

Wireless Security Systems

Wireless security systems are being used for building management. The Town Center at Boca Raton, Florida, upgraded its security systems by installing a 900-MHz Frequency Agile FA system from Inovonics. The cost of hardwire systems was almost three times more expensive.

The mall's common-area doors were equipped with magnetic contacts and glass-break detectors. When a contact trips or door glass breaks, a transmitter sends out a signal over the 902- to 928-MHz industrial-scientific-medical (ISM) frequency band to a receiver at an off-site security company. The same alarm is transmitted over the security staff's radio system identifying the location and number of the door being tampered with. The alarm also sets off a siren and strobe light at the door. When a door alarm is tripped, security personnel can get to the exact location of an alarm in approximately one minute.

The 900-MHz system includes transmitters, repeaters, and receivers. It uses frequency hopping, spread-spectrum (FHSS) technology to transmit messages. When a door alarm is tripped, the transmitters send the alarm signal 24 times on 24 different frequencies in a 10-MHz band. By spreading the information across the 10-MHz band, the system can share spectrum with other systems without experiencing interference problems.

A 64-channel receiver with 16 outputs is used. It can handle information from up to 64 transmitters and can zone received signals into 16 inputs.

Coverage of the 1.3-million-square-foot facility is accomplished by using 18 repeaters in the system. Metal structures as well as neon and fluorescent lights can cause RF noise and cancel some of the transmissions between the transmitters and receivers/repeaters. This problem was cured by repositioning the receivers and repeaters.

Wireless Monitoring

Wireless technology has also experienced rapid growth in remote monitoring, making it easier to remotely monitor wells, tanks, pumps and oil and gas lines. Mountain Water Company of Missoula, MT, is one example. Mountain Water provides the water-distribution system for approximately 50,000 people. There are 26 remote wells, tanks, and booster stations over an area of 100 square miles.

Mountain Water uses a supervisory-control and data-acquisition system (SCADA) system to control pump levels and tank levels, turn pumps on and off, switch pumps from automatic to manual, evaluate pressure and flow data, and analyze chlorine levels. Wireless technology beat out wired approaches on two fronts. Many of the remote sites do not have access to wired communications and these remote cables tend to be cut.

Traditionally, SCADA applications have been constrained to very-small-aperture-terminal (VSAT) satellite communications systems. With VSATs, there are large initial equipment costs as well as monthly fees for using the satellite network. The wireless modem system puts all of the cost up front. Once the modem is installed, you own the radio network.

Mountain Water installed EST's ESTeem model 95 modem which operates in the 72- to 73- or 75- to 765-MHz very high frequency (VHF) bands. It uses 9600-b/s RF data rate and a 20-kHz channel spacing.

A peer-to-peer configuration is used where the modem also functions as a repeater. Any station that can hear a transmission receives the transmission. If this signal has the modem's address, it processes the signal. If there is a repeat string attached, the modem passes it off along the system to the next modem. There is no need for stand-alone repeaters.

Evolving Connectivity

For about 15 years, the typical mouse for a personal computer was connected to the computer serial port and communicated one way. The mouse measured signals from an opto-coupler, made a few calculations to smooth the motion, and sent three bytes of data over the line. A microcontroller was integrated into the mouse design to allow communications via software.

Under the Universal Serial Bus (USB) the mouse is classified as a Human Interface Device (HID) and must conform to a unified command set for software configuration and control. This communication is bidirectional. There must be a response to commands or the mouse becomes disabled.

The mouse identifies itself with a unique identification code so it can accommodate multiple point devices connected to a single system. The mouse should also support error detection and recovery for the reliable delivery of data. This fully connected protocol requires that the software be able to handle up to 35 errors between the mouse and the host driver software.

System connectivity requires the ability to automatically configure the system for the user. If boxes are added or removed, the system must reconfigure itself without losing data during operation.

The local device control must be accurately maintained while it responds to the network. Error detection and correction on the network become mandatory to ensure reliable data delivery.

The HVAC control system is required to generate closed-loop control and deal with network demands. One solution is to add a second processor dedicated to the network interface. This interface must be defined carefully to avoid limiting the data flow through the system. Limited data flow means the control is compromised.

Traditional controllers are light on performance. While this may have worked fine for simple designs, the network interface-connectivity changes everything. Good logical computational engines for the finite state machines used in most network protocols and numeric capability are needed for closed-loop control calculations.

Reconfigurable Computers

New techniques are emerging from the study of reconfigurable computing which was originally proposed in the late 1960s at UCLA. Reconfigurable computing had been delayed by a lack of reconfigurable hardware. Reprogrammable logic chips like field programmable gate arrays (FPGAs) have been available for years, but these chips have only recently reached gate densities that make them suitable for more complex applications. FPGAs now have several hundred thousand reprogrammable logic gates with an anticipated doubling of gate densities every 18 months.

TSI TelSys are using reconfigurable computing technologies in ground-station equipment for satellite communications. This application involves high-rate communications, signal processing, and a variety of network protocols and data formats.

The hardware has a great deal of functional flexibility since the logic in the FPGA can be changed when necessary. This allows hardware bug fixes and upgrades to be administered as freely as software.

In order to support a new version of a network protocol, you redesign the internal logic of the FPGA. Reconfigurable computing also allows manipulation of the logic in the FPGA at run-time. The FPGA acts as an execution engine for the different hardware functions. Some are executing in parallel and others in serial fashion. The FPGA functions as a reconfigurable processing unit (RPU).

Reconfigurable computing allows the system to execute more hardware than there are gates for. This works when there are parts of the hardware that are occasionally idle. One potential application is smart cellular phones that support multiple communication and data protocols, one at a time. As the phone passes from a geographic region that is served by one protocol into a region that is served by another, the hardware is automatically reconfigured. This is the reconfigurable computing approach.

Reconfigurable computing makes it feasible to achieve greater functionality with simpler hardware. The cost of additional features is reduced to the cost of the memory required to store the logic needed.

Reconfigurable computing has the ability to execute larger hardware designs with fewer gates with the flexibility of a software-based solution and the execution speed of a more traditional, hardware-based approach. Reconfigurable computing allows the addition of new features in the field and rapid implementation of new standards and protocols on an as-needed basis.

In the multiprotocol cellular phone, it is possible to support as many protocols as can fit into the available on-board memory. New protocols could even be uploaded from a base station to the handheld phone on an as-needed basis, requiring no additional memory.

Since systems based on reconfigurable computing are upgradable in the field, such changes extend the useful life of the system. Reconfigurable computing eliminates a large amount of development effort. The logic design remains flexible and allows an incremental design flow.

Reconfigurable Hardware

Traditional FPGAs are configurable, but they are not run-time reconfigurable. Older FPGAs read their configuration from a serial EEPROM, one bit at a time. This means that the FPGA must be reprogrammed in its entirety. This is compatible with configurable computing applications, but it is not adequate for reconfigurable computing.

In order to have run-time reconfiguration, the FPGAs need to have on-the-fly reprogrammability. Partial reprogrammability is the ability to leave most of the internal logic in place and change just one part. Atmel 40K and Xilinx 62xx series FPGAs have this feature. Any gate or set of gates may be changed without affecting the state of the others. An externally visible internal state means that it is possible to capture that state and save it for later use. Xilinx 62xx series FPGAs use a 32-bit data bus called the FastMAP processor interface. It allows the internal state of the FPGA to be read and written like memory and makes it possible to swap logic designs in much the same way that pages of virtual memory are swapped into and out of physical memory.

Hardware Objects

A hardware object is a functional or logical hardware component that contains its own configuration and state information. It is a piece of logic that can be executed in an RPU. Hardware objects are relocatable and can be executed from any position in the chip.

These relocatable logic blocks have some constraints based on the hardware page size. Page sizes of 4 and 16 gates are used in the Xilinx 62xx series of FPGAs because of the routing resources at the intersections. This makes routing between hardware objects or a hardware object and its I/O pins much easier. By standardizing the interfaces of hardware objects, it is possible to maintain libraries of frequently used objects and to quickly build larger designs from these smaller components.

Hardware objects that interface to the world outside the RPU do so through an abstraction. This abstraction is called a hardware object framework which is a ring of logic in the RPU that is physically located along the outer edges. This provides a set of standard interfaces to memory and peripheral devices outside of the RPU.

In reconfigurable computing the software decides which hardware objects to execute and when to swap hardware objects into and out of the reconfigurable logic. It also performs the routing between hardware objects or between hardware objects and the hardware object framework.

The software that does this is called the run-time environment and is similar to the operating system that manages the execution of multiple tasks in personal computers. Hardware objects have priorities and it is the job of the run-time environment to organize this information and make decisions based upon it.

The run-time environment is needed to make decisions while the system is running. The input data and the appropriate logic block are handed over to the run-time environment which executes the attached hardware object.

The run-time environment must first locate space within the RPU that is large enough to execute the hardware object. It then performs the necessary routing between the hardware object's inputs and outputs and the blocks of memory needed. Commercial run-time environments for RPUs will emerge if reconfigurable computing becomes popular.

The run-time environment can be thought of as a series of three layers. The device abstraction layer is the lowest level and is analogous to the parts of an operating system that must be written in assembly language because they are processor-specific. The device abstraction layer is concerned with the hardware page size and the routing resources that are available at the edge of each hardware page. The device abstraction layer also provides the read/write interface for the layer above.

The middle layer controls the placement and routing of hardware objects. It decides where each object will be physically located within the device and adds routing between hardware objects.

The upper layer is called the object scheduler. It provides an application programming interface (API) for the RPUs and is responsible for deciding which hardware objects are currently running. This decision can be based on a scheduling algorithm such as first-come, first-served.

References

Barr, Michael, "A Reconfigurable Computing Primer," *Multimedia Systems Design*, Vol. 2 No. 9, September 1998, pp. 44-47.

Buderi, Robert, "Sleep with a Computer," *Upside*, Vol. 118 No. 1, February 2000, p. 202.

Crowley, Susan I., "Hello to our Future," *AARP Bulletin*, Vol. 41 No. 1, January 2000, pp. 3, 14-15, 20.

Keenan, Robert, "Modem Monitors SCADA Systems," *Wireless Systems Designs*, Vol. 2 No. 5, May 1997, p. 30.

Keenan, Robert, "Wireless System Secures Shopping Malls," *Wireless Systems Design*, Vol. 2 No. 4, April 1997, p. 30.

Liptak, Bela A., Editor-in-chief, *Instrument Engineers' Handbook*, 3rd

Edition, Chilton Book Company: Radnor, PA, 1995.

"The Connectivity Revolution," *Wireless Systems Design*, Vol. 3 No. 9, September 1998, pp. 1-3.

Redman, Jim, "Upgrades Enhance Capability," *Industrial Computing*, Vol. 18 No. 5, April 1999, pp. 22-25.

TigerDirect.com Catalog, Vol. X Issue 1, 2000, p. 107.

Internet: www.home.cnet.com/specialreports/0-6014-7-818759.html, Neumann, Alicia and Kristina Blachere, "Nanotechnology Will Change the World," November 1999, pp. 1-2.

Internet: www.foresight.org/EOC/EOC.Chapter_1.html, Ulmer, Kevin, "Engines of Creation," November 1999, pp. 1-10.

Chapter 4

Advanced HVAC Control, Information Technology And Open Systems

Clean rooms represent space areas that are used for testing/analysis laboratories, in medical, military and electronic processing industries. The optimization of clean room controls is an important area in the future of HVAC controls. Many of these concepts will be finding their way into building automation.

The production of semiconductors must take place in a clean room environment. Optimization of the clean room control system can reduce the cost of operation and the number of product defects. In semiconductor manufacturing, an area of 100,000 square feet (9290 square meters) can have a daily production worth of over $1 million.

Semiconductor Manufacturing

The overall semiconductor fabrication process consists of the patterning of a sequence of successive layers. The patterning steps are transferred into layers of the final system. The sequence of layers is used to build up an integrated system.

One process is the creation of a silicon dioxide insulating layer on the surface of a silicon wafer and the selective removal of sections of the insulating layer. This step begins with a bare polished silicon wafer. The wafer is exposed to oxygen in a high-temperature furnace to grow a uniform layer of silicon dioxide on its surface. After the wafer is cooled, it is coated with a thin film of organic resist material.

The resist is dried and baked and the wafer is ready to begin the patterning step.

The pattern to be transferred to the wafer surface exists as a mask. The mask is a transparent support material coated with a thin layer of opaque material. Certain portions of the opaque material are removed, leaving opaque material on the mask in the precise pattern required on the silicon surface.

The mask, with the desired pattern engraved upon it, is brought face down into close proximity with the wafer surface. The dark areas of opaque material on the surface of the mask are the areas where it is desired to leave silicon dioxide on the surface of the silicon. Openings in the mask correspond to areas where it is desired to remove silicon dioxide from the silicon surface.

The mask is laid on the wafer and its back surface is flooded with ionizing radiation, such as ultraviolet light or low-energy X-rays. The radiation is blocked in areas where the mask has opaque material on its surface. Where there is no opaque material on the mask surface, the ionizing radiation passes through into the resist, the silicon dioxide, and silicon.

The ionizing radiation has little effect on the silicon dioxide and silicon but it breaks down the molecular structure of the resist into smaller molecules. This step makes it possible to dissolve exposed resist material in solvents that will not dissolve the unexposed resist material.

The resist is developed by immersing the silicon wafer in a solvent. The pattern originally existing as a set of opaque geometries on the mask surface is transferred as a corresponding pattern into the resist material on the surface of the silicon dioxide.

The same pattern is transferred to the silicon dioxide by exposing the wafer to a material that will etch silicon dioxide but will not attack the organic resist material or the silicon wafer surface. The etching step is usually done with hydrofluoric acid, which dissolves silicon dioxide, but is incapable of etching the surface of silicon.

The final step in patterning is removal of the remaining organic resist material. The techniques used to remove resist materials involve acids, such as chromic acid and atomic oxygen, to oxidize the organic materials. The pattern then has been transferred into the silicon dioxide on the wafer surface. A similar sequence of steps is used to selectively pattern each of the layers. The steps differ in the types of etchants used.

Integrated Circuit Fabrication

A sequence of patterned layers is used to build NMOS integrated circuits and systems. The fabrication of a basic inverter circuit takes place in a system and all other circuits are simultaneously implemented by the same process.

The first mask includes the sources and drains of all transistors in the circuit, together with the transistor gate areas and any circuit interconnection paths. This mask is used for the first step in the process.

The next step is to differentiate transistors that are normally on (depletion mode) from those that are normally off (enhancement mode). This is done by an overcoat of the wafer with resist material, exposing the resist material through openings in another mask, and developing it. The pattern leaves an opening in the resist material over the area to be turned into depletion mode transistors.

The actual conversion of the underlying silicon is done by implanting ions of arsenic or antimony into the silicon surface. The resist material acts to prevent the ions from reaching the silicon surface. Ions are only implanted in the silicon area free of resist which causes a slight n-type conductivity in the underlying silicon.

The wafer is then heated while exposed to oxygen, to grow a thin layer of silicon dioxide over its entire surface. It is then coated with a thin layer of polycrystalline silicon. This layer forms the gates of all the transistors in the circuit and also serves as a second layer for circuit interconnections.

The n-type regions are diffused into the p-type silicon substrate, forming the sources and drains of the transistors and the first level of interconnections. This is done by first removing the thin gate oxide in the areas not covered by the polysilicon.

The wafer is then exposed to n-type impurities such as arsenic, antimony, or phosphorous at high temperature which allows these impurities to convert the exposed underlying silicon to n-type material. The polysilicon area and the thin oxide under it act to prevent impurities from diffusing into the underlying silicon.

Interconnections are made with a metal layer that contacts both the diffused areas and the polycrystalline areas. Another layer of insulating oxide is coated over the entire circuit and then removed in places where contacts are desired. The wafer surface is usually coated with another layer of oxide. This is called overglassing and provides physical protection for the device. Another mask is then used for the metal wire-bonding pads.

Each wafer contains many individual chips. The chips are separated by scribing the wafer surface with a diamond scribe and then fracturing the wafer along the scribe lines. Each individual chip is then cemented in a package and wire leads are bonded to the metal contact pads on the chip and to pads in the package that connect with its external pins. A cover is then cemented onto the package.

Of the large number of individual integrated system chips fabricated on a single silicon wafer, only a fraction will be completely functional. Flaws in the masks, dust particles on the wafer surface and defects cause some devices to be less than perfect. Typical conditions in the processing area are shown in Table 4-1.

Table 4-1. Clean Room Environment

No drafts	
	+0.02 inches H_2O ± 0.005 inches H_2O, or 5 Pa ± 1.3 Pa
No temperature gradients	
	72°F ± 1°F, or 22°C ± 0.6°C
No humidity gradients	
	35% RH ± 3%
No airflow variations	
	60 air changes/hour ± 5%

One goal of control optimization is to maximize productivity by the accurate control of these parameters. Another goal is to conserve energy. The control elements and control loop configuration needed for high-productivity and low energy costs are the subject of the next section.

Air Contamination

In order to prevent contamination by air infiltration from the surrounding areas, the clean room pressure must be higher than the rest of the building. The clean room is surrounded by a perimeter corridor (Figure 4-1). The pressure in the clean room is kept higher than the surrounding corridor to block the leakage of contaminated air. The clean room is made up of work areas which are considered as zones. There may be several hundred zones in a plant.

Zone Control

Air is supplied to the zones through filters located in the ceiling. The flow and temperature of the air are both controlled. The exhaust air is evacuated in the lower area of the zone. In semiconductor manufacturing, each workstation has its own air supply filter and exhaust outlet.

In order to contain any toxic fumes, clean air enters the workstation at a velocity of about 75-fpm (0.38 m/s). The air that is not pulled in by the exhaust system is recirculated back into the workstation with a fan. The rest of the air is returned by a return air header. A damper in this header is modulated to control the air pressure.

Pressure Control

The pressure in the isolating corridor acts as the reference for the area pressure controller. This controller maintains a few hundredths of an inch of positive pressure relative to the isolating corridor.

The quality of the building construction determines how high this setpoint will be. Even with lower quality buildings where leakage may

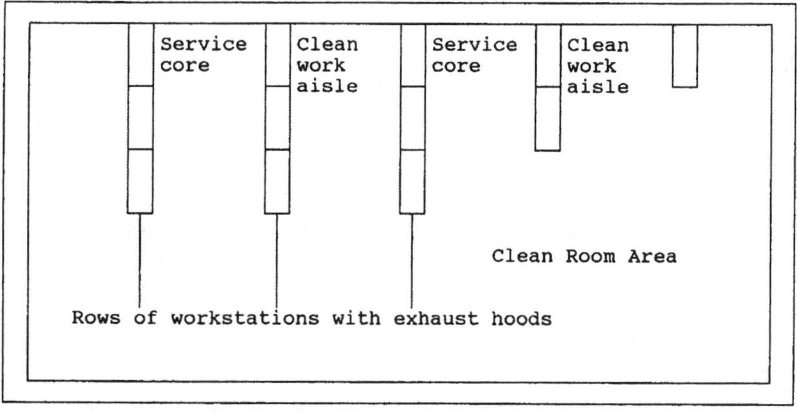

Figure 4-1. Clean room area with workstations and isolating corridor.

be high, a setting of about 0.02 inches H_2O can be maintained.

At these near-atmospheric pressures, the air behaves as if it were incompressible and pressure control is fast and stable. When the loop is energized, the pressure controller moves the return air control damper until the preset differential is reached. Then, the electric motor stops and the damper stays at its last opening. This position is not changed as long as the following conditions apply:

Return Airflow = Supply Airflow
– (Exhaust Airflow + Pressurization Loss)

When this airflow balance is affected by a change in airflow or pressure, it causes a change in the space pressure and the return air control damper responds by modifying the opening of the return air control damper.

Drafts

Productivity and energy costs are maximized if drafts are minimal in the work area. Drafts can stir up the dust in this area, could contaminate the product and reduce production. In order to eliminate drafts, the pressure at each of the work areas must be at the same value. This eliminates any pressure differential between areas and prevents drafts.

A pressure controller at each workstation will result in a uniform pressure profile throughout the area. All areas should be at a positive pressure.

The local fans will cause a lower pressure where they draw air in. These localized vacuum zones could cause contamination by allowing air infiltration. Pressure control of the vacuum zones could cause drafts from the pressure gradients.

One technique involves raising the setpoint of the pressure controller until all areas are at a positive pressure. The result depends on the quality of building construction which might not be high enough to allow operation at the higher space pressures. The pressurization loss in badly sealed buildings could make it difficult to reach the higher space pressure.

Another technique is to use a second pressure controller, which would throttle a damper in the supply air duct. This would provide the control needed but would also increase the cost of the control system by adding another control loop for each workstation.

A more economical solution would be a hand-operated bypass damper which is manually set during the initial setup and balancing. This solution is workable, but the setting would need to be changed as the manufacturing process changes.

Temperature Control

The temperature at each workstation is controlled by a thermostat. The temperature controller adjusts the ratio of cold air to hot air within the supply air mixing box to maintain the space temperature.

The temperature gradients in the clean room area should be ±1°F of 72°F (-0.6°C of 22°C). Conventional thermostats cannot generally meet this requirement because of their measurement accuracy and control quality. Even if individual calibration is used, the error in overall loop performance would still be about ±2 or ±3°F (±1°C or ±1.7°C).

Part of the error will be due to the offset that cannot be eliminated in proportional controllers, such as thermostats. In operation the thermostat must move its output from the midscale value with 50% mixing of cold and hot air. An error in room temperature must exist which is the permanent offset. The size of this offset error for the control of maximum cooling can be estimated by dividing the spring range of the cold air damper by twice the thermostat gain. Typical dampers have an 8 to 13 PSIG (55 to 90 kPa) spring which gives a spring range of 5 PSI, or 34.5 kPa. If the thermostat gain is 2.5 PSI/°F (31 kPa/°C), then the offset error is 1°F (0.6°C).

This means the space temperature must rise to 73°F (22.8°C) before the damper can be fully opened. The offset error will tend to increase as the spring range increases or as the thermostat gain decreased.

Control of the clean room temperature to ±1°F (±0.6°C) requires an RTD-sensor or a semiconductor temperature sensor with a proportional-plus-integral controller. This will eliminate the offset error and can best be done with microprocessor-based controllers that communicate with the sensors over a pair of twisted pair wires that serve as a data highway.

High accuracy temperature transmitters and sensors may cost a little more initially, but can save time and money over the long term by improving efficiency and reducing maintenance costs. If the accuracy can be extended to the entire system, the savings may be dramatic.

Direct wiring transmits the low-level sensor output over sensor extension wires. Sensor extension wires are fragile and cost about three times more than common shielded copper wire. Transmitters amplify and condition the signal, and transmit it over a twisted wire pair.

Temperature Measurement Accuracy

Accuracy is a gauge of how much a measurement may vary from the true theoretical temperature value. Generally speaking, a high-accuracy temperature measurement is greater than ±1°C for a span of 200 degrees or less.

Temperature measurement uncertainties may be introduced by the transmitter and the sensor. Temperature transmitter inaccuracies include input/output accuracy, linearity, load and line voltage effects. The input accuracy is the uncertainty of the measurement after conversion through the transmitter's analog to digital converter. Output accuracy is the uncertainty of the output after the signal is converted from digital to analog, it includes the input accuracy.

Resolution is the smallest change that can be detected in a measurement. It is listed as a percent of span or as a number of bits. Linearity is the amount of deviation from a straight line between zero and full scale input. Deviations may be caused by electrical noise or interference and the resistance of the leads depending on the sensor type and lead length. Repeatability is the ability of the unit to generate the same output value for the same input for consecutive measurements under the same operating conditions.

In some sensor specifications linearity is used in place of accuracy. Others use an accuracy specification that includes linearity and repeatability, and assumes specified ambient temperature conditions.

RTD output can change due to temperature cycling, corrosion in lead wires, moisture, and contamination. Two and 3-wire RTDs are subject to lead wire imbalances. A 4-wire RTD can provide true lead wire compensation when it is used with a transmitter that has a 4-wire input.

Each ohm of imbalance in an RTD's lead wires results in as much as 2.5°C measurement error. Imbalances are due to terminal block corrosion, connector corrosion, extension wire splices, loose connections, lead length differences and work hardening from bending.

A 2-wire RTD does not compensate for lead wire length or resistance differences. A 3-wire RTD will compensate for lead wire length if each lead is exactly the same resistance, but will not compensate for differences in lead resistance. A 4-wire RTD that is used with a temperature transmitter that accepts a true 4-wire RTD input can compensate for unequal lead lengths and lead resistance differences. A 4-wire RTD is only slightly more than a 3-wire.

An intelligent temperature transmitter will provide a constant current source to the outer leads of a 4-wire RTD. This supplies the current to the RTD. The voltage drop is measured across the inner leads, which is a high impedance, low current voltage loop. There is virtually no current flow in the voltage loop, so there is no voltage drop. The voltage is directly proportional to RTD resistance. The lead resistance has little or no effect.

Microprocessor-based temperature transmitters have universal inputs and ranges and diagnostic capabilities that can save maintenance time. They continually monitor the sensor and send a message if a sensor wire should break or stop transmitting a signal. The fault messages on an integral display can save troubleshooting costs.

Noise Interference

Noise interference can affect accuracy. Direct wiring exposes high impedance, low-level sensor signals to the harmful effects of RFI/EMI. A temperature transmitter filters out RFI/EMI noise and converts the low-level signal to a low-impedance 4-20-mA current signal. Transmitters can be calibrated for a specific range within a sensor's overall capabilities for more accurate measurements. RTDs are nonlinear and some intelligent transmitters can use up to 128 points of linearization for precise temperature calculations.

Sensor Matching

Some intelligent transmitters provide methods of trimming the input to a particular temperature sensor. The sensor is checked at a stabilized temperature and the transmitter is programmed to capture the output of the sensor. By capturing several data points from the sensor, the transmitter can compensate for the deviations of that sensor.

Another method of calibration uses the Callendar-Van Dusen equation for approximating the resistance-to-temperature relationship of a particular sensor. Most RTD temperature sensors can be obtained with constants that are then programmed into a transmitter that uses the equation to correct for divergence.

Using a 1000 ohm RTD rather than a 100 ohm unit will give better measurement resolution, since each degree change in temperature will result in a resistance change ten times greater.

Using the digital output from a smart transmitter instead of an analog output will provide higher accuracy. This eliminates the error associated with the D/A converter.

Stability

Accuracy is the level of uncertainty of a transmitter's or sensor's output at a given time while stability is the uncertainty of a transmitter's or sensor's output over a period of time. Stability is usually specified as a percent of temperature span per year. It indicates how often the system will need calibration.

Total accuracy also depends on the equipment used to calibrate the measurement loop. Test equipment should provide a 3:1 accuracy ratio, where the test device is rated at least three times more accurate than the unit under calibration. Equipment should be traceable to an accredited metrology institute such as NIST (National Institute of Standards and Technology).

Humidity Control

The relative humidity sensors are installed in the return air stream. The relative humidity in the clean room should be within RH +3% so the humidity sensor must have an error of less than ±3% RH. The repeatability of human hair element sensors is about ±1% RH. These units can be used for clean rooms provided that they are individually calibrated for operation at or around 35% RH. Without individual calibration, they will not perform satisfactorily in this application, since their off-the-shelf error, is about +5% RH.

Hair Hygrometers

These devices operate by a change of length of organic and synthetic fibers when these are exposed to a moist atmosphere. A mechanical linkage is used to amplify the element movement for readout.

Good circulation is required and in ducts there is generally an adequate gas velocity to insure a dependable measurement. However, if the instrument is to be mounted in a room, the location should be carefully chosen. The sensing portion should be in the room or compartment where the relative humidity is to be measured. The sensor should not be mounted near doors or other openings where it will be exposed to spurious drafts. Flush mounting on a panel should be avoided because the atmosphere in the back of the panel is stagnant.

Dew Point Hygrometers

The solution conductivity type of hygrometer depends on solids condensing and absorbing minute quantities of water on their surfaces. The solution vapor pressure depends on the moisture content of the atmosphere.

The sensing element is a thin-walled, hollow-metal socket wrapped with tape impregnated with lithium chloride salt crystals. Two wires are wrapped over the tape and connected to a regulated, alternating current voltage source. The electric circuit between the wires is completed by the salt crystals.

When the sensing element is exposed to the sample atmosphere, water condensing on the crystals forms an ionic solution which permits an electric current to flow between the wires. This current in turn heats the solution and raises its vapor pressure.

As the water condenses, more current flows, which results in a further increase of the solution vapor pressure until equilibrium is reached. A decreasing moisture content will cause water to evaporate from the element, decreasing the current flow and resulting in a new equilibrium at a lower vapor pressure. A temperature sensor inside the hollow socket is used to detect the temperature.

The output can be calibrated in terms of dew point temperature water vapor pressure or specific humidity. Relative humidity is obtained by converting sample and dew point temperature readings to percent relative humidity.

A sample velocity in excess of 1 foot (0.3 m) per second can result in a poor measurement and shortened element life. At high sample velocities, convective heat losses swamp the measurement signal and the additional heating power can reduce the element life. The sensing element should be located in a relatively quiescent zone or must be protected from direct impingement of the flow. In ducts, a sheet metal hood can be installed over the element. It is open on the downstream side. In piping, the element can be installed in the side outlet of a tee or mounted separately in a sampling chamber. This sensor requires periodic maintenance and must be removable.

Surface Conductivity Hygrometers

Every object in a moist atmosphere has water molecules on its surface. The concentration on these molecules is related to the temperature of the object and the dew point of the atmosphere. If the temperature of the surface is above the dew point of the atmosphere, the thin layer of molecules is not visible. As the surface cools to the dew point, the density of water molecules at the surface becomes so great that water condenses on the surface and dew can be seen.

At surface temperatures above the dew point, the moisture density at the surface can be detected electrically although the water vapor is not visible to the eye. This water vapor will permit a current to flow on the surface of an insulator. This current flow depends on the moisture density at the surface.

The measuring element consists of a polished inert surface with a metal grid and a thermocouple imbedded in the surface. A fixed potential is maintained across the grid and the current flow is compared to the reference current flow at dew point.

This signal is amplified and used to modulate a cooler so that the surface is maintained at the dew point of the sample. The cooler is a crystal that pumps heat away from the sensor when electric power is supplied to it. From the thermocouple imbedded in the surface, the output can be presented as dew point temperature or in terms of relative or specific humidity.

Chilled Mirror Hygrometers

The optical, chilled mirror dew point technique is a fundamental measurement of the humidity of a gas since the saturation temperature determines the saturation partial pressure of the water vapor. This relationship is used by the National Bureau of Standards for humidity sensor calibration. The saturation vapor pressures over plane surfaces of pure water and pure ice correspond to equivalent dew points or frost points.

The equilibrium partial pressure is illustrated in Figure 4-2. At the saturation temperature, or dew point, the gas mixture is saturated with respect to water or ice and the rate of water molecules leaving the atmosphere and condensing on the chilled surface is the same as the rate of water molecules leaving the chilled surface and reentering the atmosphere. At equilibrium saturation, the water vapor partial pressure of the condensate is equal to the water vapor partial pressure of the gas atmosphere.

This equilibrium at the mirror surface is obtained by cooling the mirror to the saturation temperature. A temperature sensor is in direct thermal contact with the mirror, and the mirror temperature is used indirectly as the dew point or saturation temperature.

The cooling of the mirror surface may be accomplished with mechanical refrigeration. More recent units use thermoelectric heat pumps. Detection of the condensation is done with optical phototransistor detection. The newer units automatically control the surface at the dew point or frost point.

Manually cooled, visually observed hygrometers are known as dew cups. These devices have been replaced by the thermoelectrically cooled, optically observed dew point hygrometer for on-line measurement. Here, the mirror surface is chilled to the dew point by a thermo-

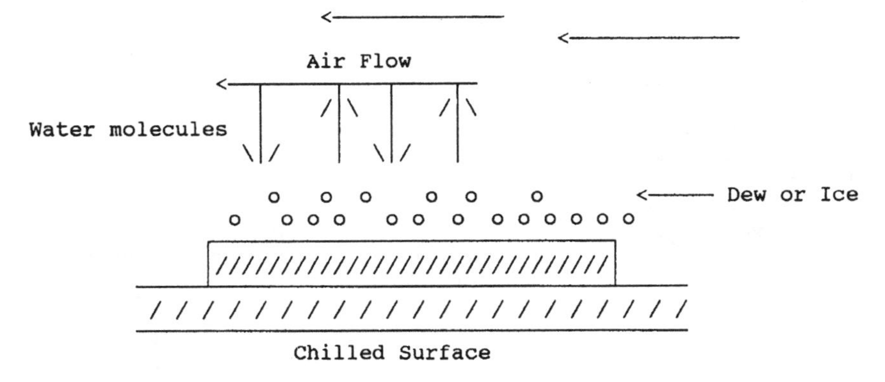

Figure 4-2. Equilibrium partial pressure.

electric cooler while a continuous sample of the atmosphere passes over the mirror. The mirror is illuminated by a light source and observed by a photodetector bridge network.

As condensate forms on the mirror, the change in reflectance is detected by a reduction in the reflected light level received by the photodetector because of the light-scattering effect of the individual dew molecules. This light reduction forces the optical bridge toward a balance point, reduces the input error signal to the amplifier, and proportionally controls the drive from the power supply to the thermoelectric cooler (Figure 4-3). This maintains the mirror at a temperature at which a constant-thickness dew layer is retained. Embedded in the mirror, a temperature-measuring element measures the dew point temperature. Advances in the technology of thermoelectric coolers allow practical instruments with several stages of thermoelectric cooling for most dew point ranges.

Relative humidity is controlled at the zone level. The control action is based on the relative humidity reading in the combined return air stream from the workstations in that zone.

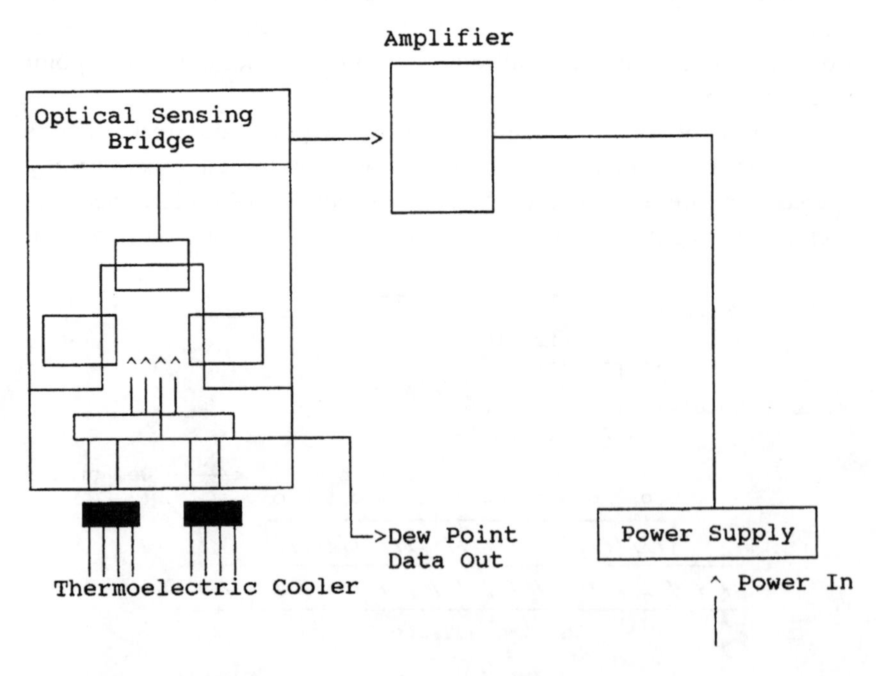

Figure 4-3. Chilled mirror hygrometer.

Flow Control

Each workstation has its own mixing box for modulating the ratio of cold and hot air to satisfy the requirements of the space thermostat (Figure 4-4). The total air supply flow to a workstation should be equal to 60 air changes per hour plus the exhaust rate from that subzone.

The total air supply rate should be controlled to ±5% of the actual flow over a flow range of 3:1. The rangeability of 3:1 is needed to sup-

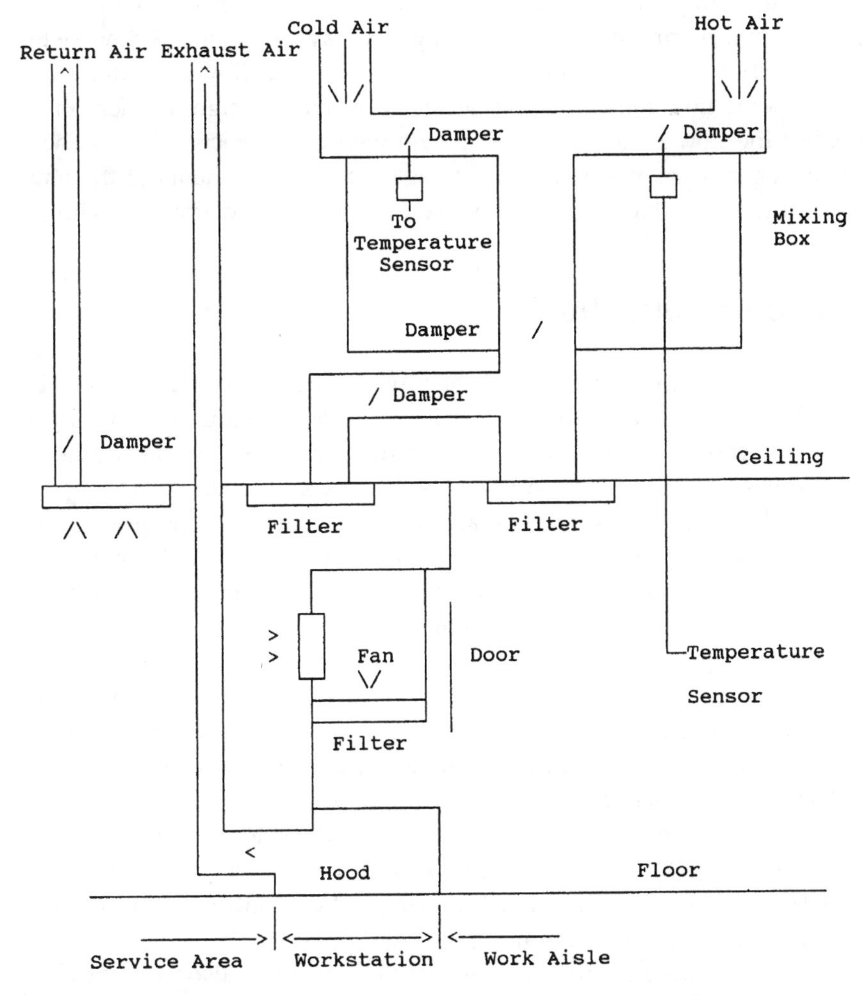

Figure 4-4. Workstation environment control.

port several semiconductor manufacturing processes with their associated exhaust requirements.

If the flow controller is set manually, it must be changed each time the process step changes. The setting of the flow controller needs to be done with an individual in-place calibration using a portable hot wire anemometer reference.

Some of the available mixing box designs should not be used for this application. Some of these mixing boxes are pressure dependent where the total flow change has some dependence on the air supply pressure. Pressure-independent types need to be used since both the cold and the hot air supply pressures to the mixing box may vary some. A flow meter with a 3:1 rangeability and accuracy of ±5% of actual flow is needed.

Some flow meters use a selector or override design. In these units either the flow or temperature is controlled on a selective basis. These override designs will periodically neglect the requirements of the temperature controller and will produce upsets and temperature cycling.

Zone Optimization

Each row of workstations makes up a zone. Each zone is served by a cold deck (CD), a hot deck (HD), and a return air subheader or finger. The operation of the controllers for the individual workstations or subzones are dependent on the zone finger conditions.

The subheader should be able to control the subzone pressure as long as the pressure difference across the damper is high enough to remove all the return air without forcing the damper to full open. As long as the dampers are throttling and not fully open or completely closed, the pressure controller is in control mode.

A nonlinear controller is used to control the vacuum in the return air finger. The nonlinear controller has a dead band that protects the cold deck finger temperature from being changed until a substantial change takes place in the return air pressure.

The mixing box will be in a control mode as long as its damper is not forced to a full open or closed position. If a damper is fully open, the control loop is out of control. Damper position controllers are used to prevent the dampers from moving to full open.

The relative humidity in the return air must also be controlled within acceptable limits. Figure 4-5 shows how this can be done.

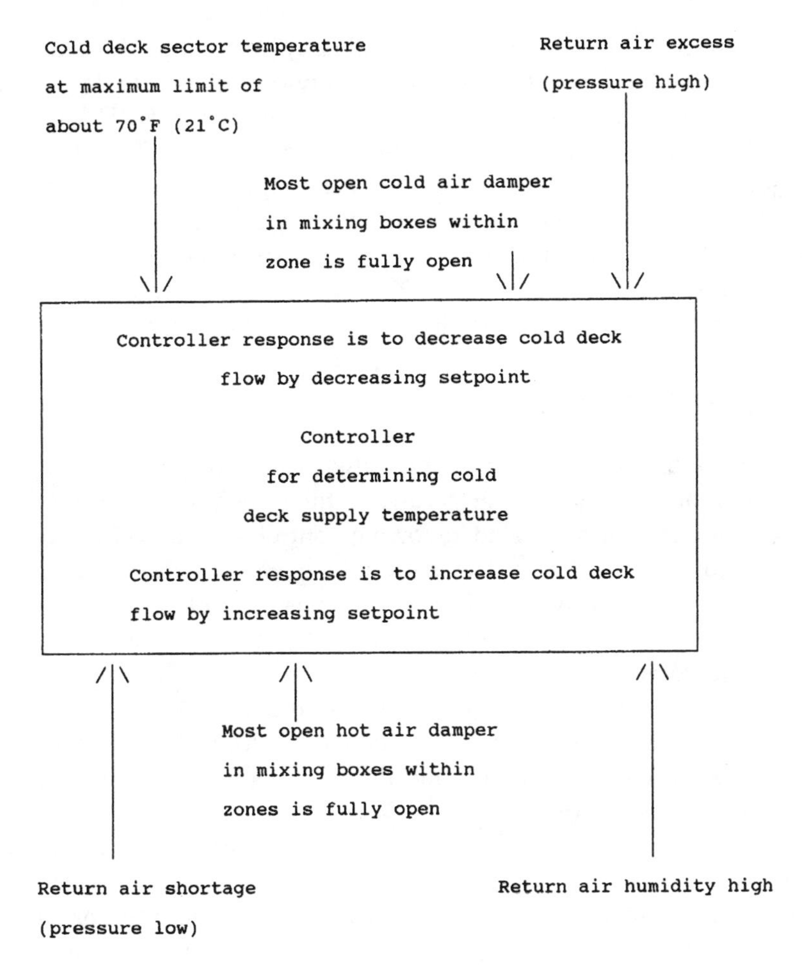

Figure 4-5. Control for cold deck temperature optimization.

Envelope Control

When the number of control variables is greater than the number of manipulated variables, multivariable envelope control can be used. In this case the manipulated variable is the temperature setpoint. It is selectively controlled to keep several variables under control.

The setpoint can be used to change the cooling capacity of each unit of cold deck air. The same cooling can be done by using less air at lower temperature or by using more air at higher temperature.

Return air humidity can be controlled by modulating the temperature setpoint. Increasing the ratio of humidity-controlled cold deck air in the zone supply brings the zone closer to the desired 35% RH setpoint.

In this control system, a temperature setpoint will increase the cold deck demand which lowers the hot deck and return air demand. A temperature setpoint decrease will occur when a return air excess is detected or the cold air damper in the mixing box is fully open. If the cold deck finger temperature exceeds 70°F (21°C), the temperature setpoint will decrease to keep the cold deck cooler than the hot deck.

Plantwide Control

If the temperature and the humidity of the cold deck supply air are fixed, and if a damper or a valve is fully open or closed, the system is out of control. A load following control system will automatically adjust the main cold deck supply temperature and the humidity. This prevents the valves or dampers to cause lost of control by fully opening or closing. The result is increased productivity and reduced operating costs.

A plant might consist of several dozen zones. The plant control system has a hierarchical structure (Figure 4-6). The workstation controls are subject to the zone controls and the plant controls check the operation of the zone controllers.

An important action of the levels of the hierarchy is the active control of valve and damper positions. When a control valve or a damper is approaching the point of losing control, (near full or closed position) the load-following control system at the next higher level modifies the air or water supply conditions.

The plant control system acts as a flexible set of material balance and heat/humidity balance controls. Load-following optimization of the setpoints produces heat balance control. If the material balance requires an increase in airflow and the heat balance requires a reduction in the heat input to the space, both requirements will be met by admitting more air at a lower temperature.

Material Balance Control

The plant material balance uses pressure control to modulate variable-volume fans to maintain a minimum supply pressure in the cold and hot decks. The cold deck supply fan uses outside air.

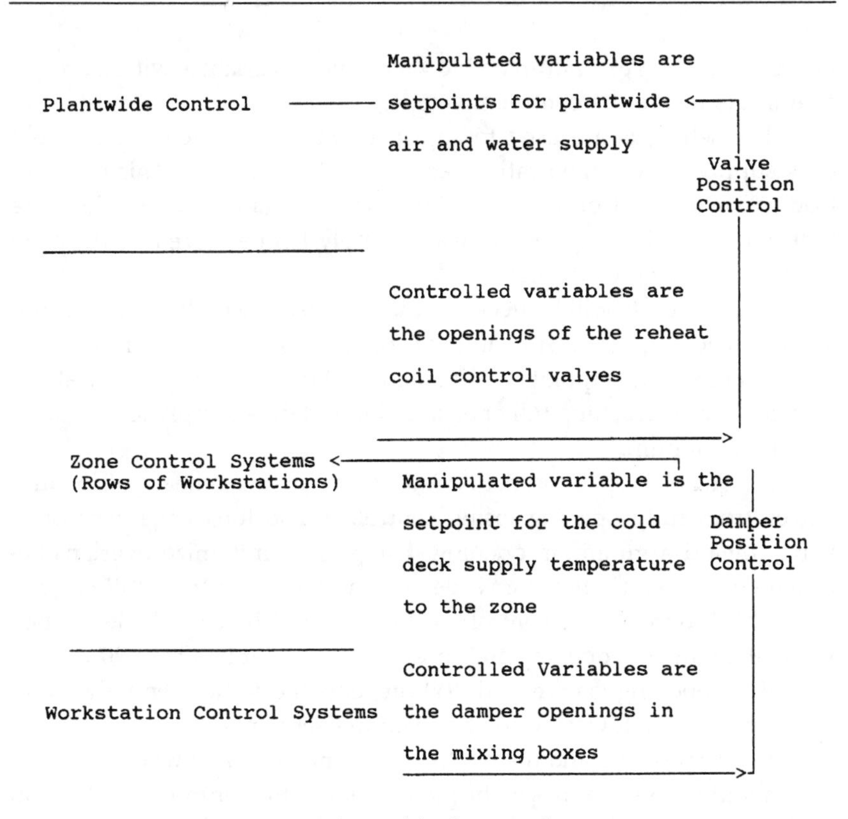

Figure 4-6. Hierarchical control structure.

The suction pressure of the hot deck supply fan is an indication of the balance between return air availability and hot deck demand. This balance is maintained at the zone level by a controller that must be tuned for slow, gradual action.

Sudden upsets or emergency conditions are corrected by a relief damper if the suction pressure is high (low vacuum) and a makeup damper if it is low (high vacuum). Between these limits, both dampers are closed and the suction pressure is allowed to float.

Heat Balance Control

Heat balance requires multi-variable envelope control since the number of controlled variables is greater than the number of available manipulated variables. The plant air and water supply temperature setpoints are selectively modulated to keep several variables within acceptable limits inside the control envelope.

The setpoints are used to adjust the cooling capacity of the cold deck supply air and the heating capacity of the hot water. This provides load-following control for the minimization of operating costs. Costs are reduced by minimizing the hot water supply temperature and the cooling and reheating of the cold deck air.

The setpoints will be decreased if the least open reheat coil control valve is reaching full closure and the cold deck supply temperature is at its maximum. The setpoints will be increased if the most open reheat coil control valve is reaching full open and the cold deck supply temperature is at its minimum.

A control loop is used to prevent any of the reheat coil control valves from fully opening losing control. It also forces the least open valve toward a minimum opening. This should minimize overlaps between cooling and reheating while keeping the valve from full closure.

This is done with a control function f(x) which keeps the least open valve at about 10% opening and most open valve at less than 90% open. If the 90% opening is exceeded, f(x) prevents the most open valve from fully opening by lowering the temperature setpoints.

Load-following and the modulation of main air and water temperatures should keep all zones in the plant under stable control provided that the loads are similar in each zone. Control will be lost if the design limits are reached or if the load distribution is greater than anticipated. In this case, one zone may require large amounts of cooling (reheat coil valve closed) while another zone requires its reheat coil valve to be fully open. The control system must decide which condition it must correct, since it cannot correct both. The control system must be configured to give priority to preventing the reheat valve from fully opening.

If the difference between the opening of the most open and the least open valves should reach 95%, a valve position alarm should be actuated. This alarm would indicate excessive loads between zones and allow corrective action to be taken by revising the system which would include modifying air supply ducts and adding or removing mixing boxes.

Controlling the reheat valves from being nearly closed will eliminate unstable (cycling) valve operation. This also minimizes pumping costs by minimizing pressure losses through throttling valves and minimizes heat pump operating costs by minimizing the required hot water temperature.

Zone Humidity Control

At the zone level, the return air humidity is controlled and the dew point of the cold deck air supply is modulated. The cold deck supply temperature is also modulated to follow the load. This is done by measuring the relative humidity in the return air.

The control loop selects the return air sectors with the highest and the lowest relative humidities. The loop will herd the relative humidity readings so that the highest and the lowest will be within the control limits of 35% ±3% RH.

The highest reading is used as the controller measurement and the lowest reading as the setpoint. A humidity change in either direction can be recognized and corrected using this herding technique. The setpoint of the relative humidity controller is produced by a function generator. It prevents the most humid return air sector from exceeding 38% relative humidity and to keep the driest relative humidity sector from dropping below 32% RH. The setpoint is held at 35% as long as the driest sector reads 34% RH or more. If it drops below this value, the setpoint is raised to the limit of 38% RH in order to reverse this low humidity condition, without allowing excessive humidity in the return air of some other zone.

This control should automatically respond to seasonal changes and provide control as long as the loads in the various zones are similar. If this is not the case, some zones may be moisture-generating and others require humidification.

The low humidity condition will not be controlled and the high humidity zone will be controlled to prevent it from exceeding 38% relative humidity. The intermixing of the return air sources will transfer moisture from zones with excess humidity to zones with low humidity. When the difference between the minimum and maximum humidity reaches 8%, an alarm is actuated. This will allow adjustments to be made.

Exhaust Air Control

Exhaust air control is done with a two-position damper. When the workstation is operating the damper is open. A switch indicates when the damper is open. The minimum damper position should allow enough air exhaust flow for operator safety. Operating costs are lowered because less outside air needs to be conditioned if the exhaust airflow is lowered.

In order to maintain the required exhaust air flows accurately, the vacuum in the exhaust air ductwork must be kept at a constant value. A pressure controller keeps the vacuum constant by throttling the exhaust air damper. If these dampers are to remain in control, they must not go to full open.

This is done by identifying the most open damper and then comparing this opening with the setpoint of the damper position controller. This controller will limit the opening of the most open damper to 80% and increase the vacuum in the exhaust air header if the opening exceeds 80%.

If the pressure measurement for the controller goes below 9 PSIG (0.6 bar), the vacuum setpoint of the controller is increased (pressure setting lowered). This increases the operating level of the exhaust fan. Limits on the setpoint of the controller are used to prevent damage, such as the collapsing of ducts from excessive vacuum.

A glycol-circulating heat recovery loop can be used to preheat the entering outside air or as a heat source to a heat pump in the winter. Operating costs are lowered by recovering the heat content of the air before it is exhausted in the winter.

Pollution Control

The discharge of chemical vapors into the atmosphere is a pollution problem. The usual approach is to remove most of the chemical by adsorption and scrubbing before exhausting the air. Some additional benefits are provided by exhausting the air at high velocity. This will increase the dispersion in the atmosphere. Since the volume of air being exhausted varies, an air velocity controller is used to keep the velocity of discharge constant. This is done by modulating a variable orifice damper.

Intelligent Alarms

Recent studies suggest that U.S. industry loses more than $20 billion per year to preventable process disruptions. Globally, the impact is much greater. Abnormal situation management (ASM) focuses on systems that provide operator decision support and troubleshooting assistance during disruptions. ASM requires the management of large amounts of data to filter nuisance information, diagnose the cause of the alarm condition and implement a corrective action.

Future intelligent alarm management solutions may apply expert systems. These systems provide a framework for diagnosing process disruptions, generating more precise information and implementing corrective actions.

Expert Systems Reasoning

Expert systems are applications of artificial intelligence (AI) that apply inference engine and fuzzy-logic technologies to reason in real time about events that occur in dynamic processes. The inference engine reasons from rules defined in a knowledge base that is derived from expert specific knowledge.

Real-time expert systems can be divided into two basic categories: advisory and supervisory control. In advisory applications, the expert system reasons about changes in data, makes decisions based on events and presents conclusions and rationale to the operator.

The expert system that acts as a supervisor; it has the ability to provide information about control occurrences and impending problems. In some more robust supervisory control applications, the expert system will proactively adjust setpoints and switch discrete equipment on or off to resolve problems, optimize control and attain other objectives as defined in the knowledge base.

Alarm management expert systems can be integrated with the operator interface (OI) to supervise the control and provide on-line advisory or supervisory control functions. The expert system reasons about alarm points defined in a database according to tagged values, dynamic changes in a group of tagged values, or the state of other alarms in the database.

The use of meta alarms, which are alarms that reason about the state of other alarms, allows a higher level of logic that captures the dynamic changes occurring in the environment.

Conventional Alarm Systems

Conventional alarm systems are typically advisory in nature. They are tag-centric. They provide discrete alarms based on an absolute limit that is defined for a single processing value. They mirror the function of microprocessor-based annunciators by generating messages, annunciating the alarm conditions and logging the event with a data and time stamp.

The alarm is usually identified as HiHi, High, Low, or LoLo to advise about the alarm condition urgency. This identification does not take into account the current control state, such as start-up, normal, purge or shutdown. This results in many nuisance alarms.

Discrete alarm systems also tend to point to the symptoms of an abnormal situation instead of the cause. There is no capacity for reasoning about the state of other alarm points in the database.

Expert systems represent an advancement in alarm management. Advanced alarm management concepts include a sensitivity to process state, reasoning about alarm persistence, meta alarms, expanded alarm priority and alarm inhibition. They also provide intelligent assistance in managing abnormal situations by defining actions for alarm entry, acknowledgment, or exit.

Expert systems can provide alarm management strategies that consider dynamic environments where alarm priority is dependent on process state. For an alarm message to be generated, the entry conditions must be true and the process state must also be true. Alarm priority can be configured to temporarily mask nuisance alarms.

If a parent alarm is flagged as the root cause of the problem, its child alarms, the symptoms of the root cause, are temporarily inhibited before an alarm flood can occur. Parent-child alarm relationships can be defined based on a fault-tree relationship that identifies the hierarchical relationship between alarm points in the database.

Alarm inhibition can be applied based on persistence, which is the frequency of an alarm entry condition within a specified period. This prevents an alarm from being generated unless it occurs x times in a time period.

When additional information on an alarm is required, a graphical user interface (GUI) provides mouse button access to information about the events that caused the alarm or to an on-line manual that provides access to control strategies and corrective control procedures. This information may be available in HTML or other document formats. Distributed embedded intelligence provides improved operator advice and diagnostic information.

Expert systems go beyond advisory systems and distributed embedded intelligence by implementing actions based on conditions of alarm entry, acknowledgment, and exit. An action may be defined for each condition using a scripting language such as Visual Basic. Actions may include a text message, paging or e-mail through an integrated communications protocol such as Microsoft Exchange in the Windows NT environment.

More robust supervisory control applications allow the expert system to apply a corrective control strategy such as changing setpoints. Corrective actions may be applied upon alarm entry or after the alarm condition is acknowledged.

The migration of automation applications to Windows NT allows expert system, OI, and distributed control system (DCS) applications to provide application set-up and configuration using a GUI with familiar Windows point-and-click functions. Intelligent alarm management software can take advantage of the Windows NT environment to configure alarm points for entry and exit using predefined function blocks such as AND, OR, and NOT and drag-and-drop actions. This type of intuitive interface accelerates system acceptance while reducing time requirements and costs for implementation and maintenance.

A GUI lends itself to clearer, more effective message displays, annunciation, and information management. There is greater control over attributes of the operator display, including color conventions, fonts, persistence of alarms, and message behavior in terms of scrolling, zooming, and filtering.

Point-and-click interfaces lend themselves to quick manipulation of alarm messages and process information. This includes clicking for information about a specific alarm point or group of alarms, filtering alarms according to priority or group, and accessing other sources of information such as on-line manuals.

Expert systems provide tools for alarm management and correction of process disruptions. The migration of the process automation industry to Windows NT is enabling solutions for data integration that support a tighter integration of expert system applications with field-level instrumentation, OIs, DCSs, and supervisory control and data acquisition (SCADA) systems (Figure 4-7).

The OLE for Process Control (OPC) Foundation is supporting data integration by developing an open interface standard. Intellution, Fisher-Rosemount, and Honeywell are building systems with the OLE and the DCOM protocols, which allow software components running on

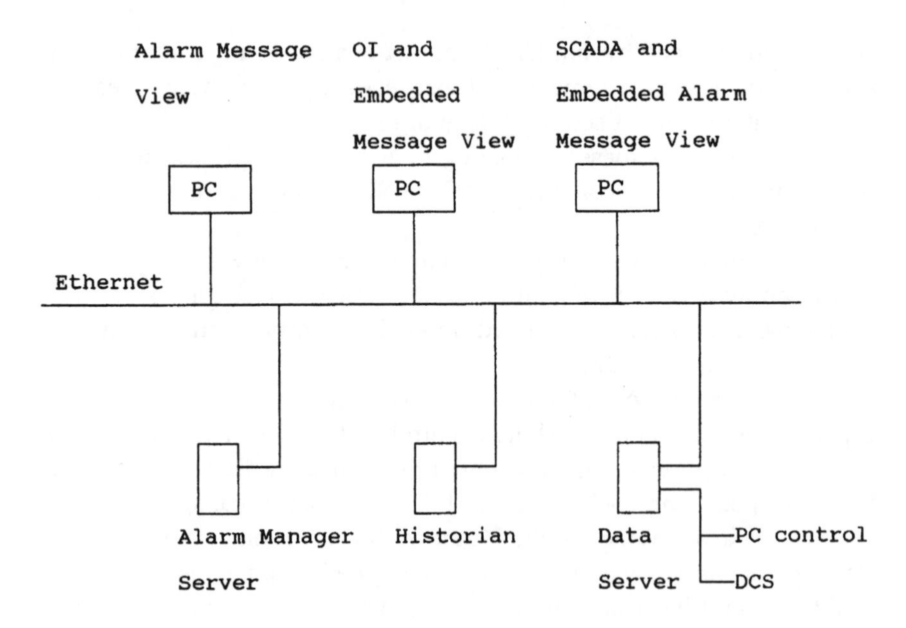

Figure 4-7. Intelligent alarm management with integrated control.

different workstations to communicate.

Intelligent alarm management systems, such as Comdale's SmartWorkX Alarm Manager, fit into an integrated control system. An integrated system can export historical data to business applications such as spreadsheets or relational databases for use in failure analysis and preventive maintenance.

The emergence of open data communication standards is eliminating the need for proprietary systems that bridge the gap between hardware and software components produced by different manufacturers. The low cost and high performance of Intel-based systems and their operating systems, such as Windows NT, are attractive to many new applications including HVAC control. This creates more flexibility for control solutions and reduces costs.

Fault Tolerant Systems

As the use of computer technology increases, so does the need for hardy systems that ensure uninterrupted operations. Several suppliers offer systems with fault tolerance. The basic way to manage a failure is to switch to an alternative component that continues to function during a failure. This redundancy is fundamental for a system that either recovers from or masks failures. Redundancy can be passive or active. The goal is to recover from failures in passive redundant systems and to mask failures so that they are transparent to the user in active redundant systems.

Passive Redundancy

A passive redundant system provides access to alternative components that are not associated with the current task and must be either activated or modified in some way to pick up the failed component's load. The transition is noticeable and may even interrupt service and degrade system performance.

Examples of passive redundant systems include standby servers and clustered systems. The mechanism for handling failures in passive redundant systems is to switch over to an alternative server. The current state of the application may be lost, and the application may need to be restarted. The restart typically causes some interruption or delay in service to the users.

Passive solutions are offered by NCR, Vinca, Microsoft and Novell and active redundant systems are offered by Stratus/Ascend and Tandem/Compaq. All require that the OS and the application have specific knowledge of the system architecture to take advantage of redundancy. This means that the OS and applications must be modified.

Systems that recover from failures use a single system to run the application until a failure occurs. The detection of a failure may take several seconds to several minutes before the recovery process begins.

In the simplest type of recovery system, an operator moves the disks from the failed system to another system and boots the second system. In more sophisticated systems, the second system has knowledge of the application and users, it reboots the applications and logs on the users. In either case there is a pause in operation and a lose of data. Applications that have been modified to know the system architecture can reboot automatically, providing a smoother recovery.

Some automatic backup systems periodically copy certain files onto another system that is then rebooted if the first system fails. The configurations are known as clusters. These cluster's include standby servers and computer arrays with fault-tolerant servers.

In systems that use standby servers (Compaq) and cluster servers (Microsoft), the storage system is divided into two parts, with each part dedicated to one of the servers. A SCSI bus between the two systems provides common access and the hard drives for both systems connected to the bus (Figure 4-8).

These simple cluster architectures provide a fail-over mechanism that lets one processor take over for another if a failure occurs. They are not transparent to the user and do not provide continuous processing.

Another type of cluster configuration uses a shared data architecture, which gives multiple computers access to the same disks. A fault-tolerant file system with universal access and a distributed lock manager are used for controlling access to files and modifying common files. These shared access clusters appeared in the early 1980s. Each server is connected through a redundant high-speed hub or switch to each other server (Figure 4-9). This configuration is used by Compaq.

Active Redundancy

An active redundant system provides an alternative processor that runs concurrently on the same task. During a failure it provides continuous service without a noticeable interruption. Because two processors are manipulating the same data simultaneously, the failure of a single component will be transparent to both the application and the user.

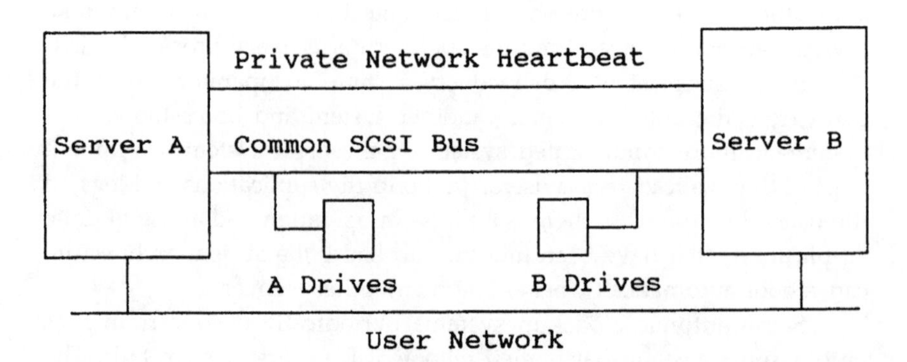

Figure 4-8. Simple cluster configuration, Microsoft cluster server.

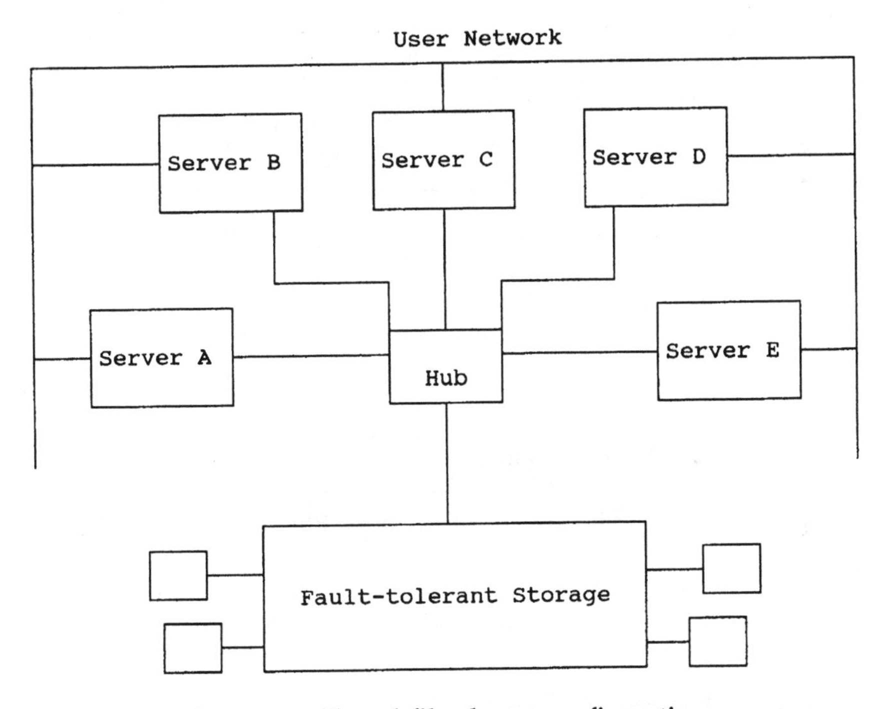

Figure 4-9. Shared-file cluster configuration.

Systems that mask failures use parallel components. Each component is capable of doing the job. If one fails, the other continues. The backup systems can be connected by some type of network connection, such as high-speed Ethernet. The data stored on the systems are mirrored over this path which also carries heartbeat messages from each system on their mutual status. This technique is used in the Vinca cluster and in the Octopus backup system.

Active redundant, fault-tolerant systems use at least two processors and custom hardware in a fail-stop configuration (Figure 4-10). Two processors are used in microcycle lockstep with hardware comparison logic to detect an error in the output of the two systems.

If the output disagrees, the system is stopped. Fail-stop systems are used by Stratus and Tandem. A special OS manages faults, errors, and system synchronization. The applications that run on these systems require modifications in order to use the fault-tolerant features.

Assured availability servers identify errors or failures before any data become corrupt. These systems isolate errors or failures so they

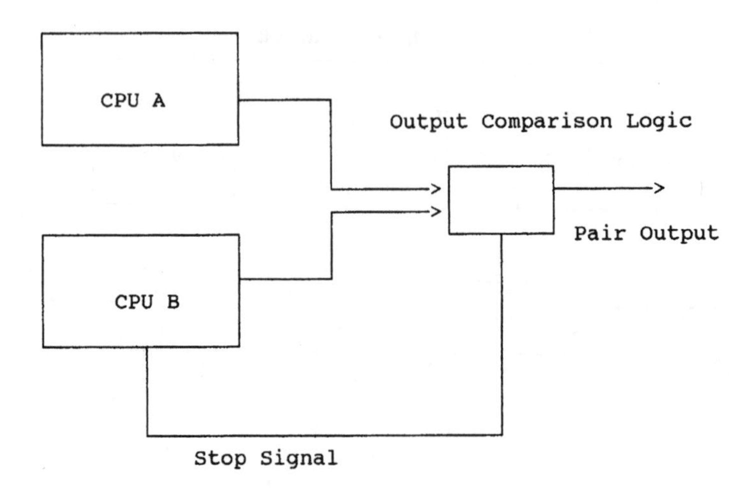

Figure 4-10. Fail-stop CPUs.

can continue to operate in the presence of the error or failure. The failed component can be repaired while the system is in operation.

Disaster-tolerant Systems

Local disaster-tolerant systems site redundant parts of the system in different locations within a building or building complex. Physical firewalls are used to separate the redundant components.

Remote disaster-tolerant systems place redundant parts of the system more than a mile apart. Commercial systems, such as the passive redundant clusters from Compaq, provide data and hot-standby capabilities over a distance of several miles.

Information Technology

The information technology (IT) revolution that has been taking place in offices is rapidly moving into building operations. This revolution is now on its way to control systems as well as sensors and actuators. The technology is robust, fast, easy to use, and ready to run buildings and plants. Users can now reap the benefits of this inexpensive yet powerful technology.

Several key enablers are pushing the technology into automation. Widespread Internet availability provides a universal information delivery mechanism. Internet and Intranet technologies provide the backbone for distributed information and control system architectures that grow with the use of object-based applications.

Objects and components play an important role supporting the distributed architecture that Internet technology provides. Object technologies let users achieve effective models of the processes under control and reduce the time needed to implement new systems. They do this by creating libraries of component models that are easy to use and reuse which allows projects to grow and prosper.

These modular, objects allow incremental improvements in the control system without having to reconfigure the complete system. In the future object-oriented tools will expose even greater benefits as the next generation of object-based control systems develops.

Mass-market Technology

Instead of using proprietary architectures, products based on mass-market business operating systems and development tools tend to leverage the large existing installed base. Many human machine interface (HMI) products rely on Visual Basic instead of their own scripting language to save costs.

The closed, proprietary nature of automation is mostly in the past. Open solutions provides easily integrated packages from different sources. Open control overcomes the limitations of proprietary systems from single sources.

Many distributed control system (DCS) suppliers provide a Windows NT-based system. Software suppliers provide soft DCS systems and control solutions based on PC-based architectures and open-networking protocols. GM has already tested NT real-time extensions to validate their performance and reliability. The new breed of open-architecture control systems is based on PC control.

NT - De facto Standard

This revolution includes Microsoft's Windows NT emergence as a de facto standard for operator interfaces and for PC-based control. Windows products are appearing in a wide range of scalable configurations, ranging from devices to plantwide data servers. Windows 2000 is designed to replace NT as a crashproof networking system. The Windows Millennium Edition (ME) should be the new personal operating system. It should have many features that were promised but lacking in Windows 98. Microsoft also has its Windows Distributed InterNet Applications architecture, which integrates Web and client/server computing.

Windows CE is evolving as the platform of choice for embedded HMIs and controllers. Microsoft's component object model (COM) and distributed component object model (DCOM) are the glue that holds systems together.

Sun Microsystems is also making an impact in the automation area with Java and the potential of software that can be written once and run anywhere. Sun supports control systems with its Java Control and Automation Framework initiative. Java-based browsers and components are also available for automation. In the embedded controller area there is EmbeddedJava and PersonalJava. There is a range of standard products with real-time extensions to Windows NT and CE, Java, and OLE for process control (OPC).

The Object Management Group (OMG) has a standard for distributed object computing called CORBA (Common Object Request Broker Architecture). CORBA competes with Microsoft's DCOM. The CORBA model was used by Foxboro to develop the Foxboro I/A Object Manager.

Another trend involves Enterprise Java Beans (EJB). Distributed-applications programming is becoming an enabler of Internet technology. All of this advances Object-oriented Remote Procedure Call (ORPC) programming which is a programming model that allows developers to easily create distributed applications.

The increased interest in distributed-application programming coincides with the growth of the Internet. Technologies such as DCOM, CORBA and EJB use the notion of location transparency.

This is an abstraction that allows a programmer to call on an object without knowing where that object actually resides. Programmers are allowed to program in the context of a familiar model. DCOM defines what an object is at a binary level.

Object-oriented programming is superior over earlier development methods since it simplifies the development of complex applications. Object-oriented programs communicate by sending messages between different objects. It takes at least one physical network round-trip to transmit a message.

The current trend of adopting Ethernet for industrial applications grows and many control applications will leverage existing desktop and networking technologies to become affordable. Automation systems must also easily interface to business systems, which are based on open standards.

The growth of open component-based systems shows the importance of open standards. A common framework allows users to choose the best combination of software modules. At the control level, this occurred when PLC suppliers added Ethernet connectivity and new open controller architectures based on common PC hardware. Other trends include Java-based control systems and the continued development of OPC interfaces.

Ethernet Use

In the area of networks, there is the trend towards using Ethernet for plant control applications. This low-cost, reliable technology brings the benefits of using Ethernet in areas beyond its traditional business applications. The total infrastructure includes Internet/Intranet networks with Ethernet at the control level.

With its speed, performance, low cost of development, and constantly updated technology, Ethernet is a good fit at the control level. It has the ability to communicate with multiple devices and manage the traffic to the information level of the plant.

New technologies allow Ethernet to be adopted in control network applications. These include low-cost Ethernet switching hubs.

The move towards more open systems includes the use of Ethernet as a control network. Fisher-Rosemount's DeltaV system uses standard Ethernet as the control network between user workstations and controllers. A redundant Ethernet capability is provided using dual-port Ethernet cards.

Foxboro was one of the first to adopt the use of the IEEE 802.3 Ethernet specification as the backbone of a control system. In its I/A series system, Foxboro first modified the protocols to optimize the IEEE 802.3 as a control bus. Then Foxboro offered standard Ethernet protocols and components as an option.

Westinghouse provides an open control system with its Ovation product. Ovation uses a scaled architecture to provide the flexibility needed for small- to very-large-scale systems. At the lower end of this scale, Ovation can use standard Ethernet products as the backbone of the control network. Other companies like GE Fanuc and Schneider Automation use Ethernet to network PLCs. Foundation Fieldbus can be adapted to use Ethernet and the Internet protocols.

Ethernet Benefits

Ethernet TCP/IP is a network technology that is sold in high volume all over the world. The growing acceptance of standard hardware and the Windows NT operating system increases Ethernet's chances of becoming the backbone of many future control and device networks.

Ethernet is already compatible with existing industry-standard PC hardware and operating systems. Many PCs include Ethernet interfaces. Windows NT is becoming the industry standard operating system for PCs and workstations and it has built-in Ethernet drivers.

Ethernet TCP/IP provides connectivity to the Internet and is moving into more applications in automation and control. Devices on an Ethernet TCP/IP network need only to be assigned an IP address for Internet connectivity.

Distributed applications must contend with an abundance of failure possibilities. In contrast to older applications that run a single process on a single computer, a distributed application has multiple potential points of failure. Distributed applications must contend with network brownouts, server crashes and operating system failures. The common point of failure becomes the transmission and receipt of messages between the components of the distributed application.

An operating system is needed that facilitates the reliable transmission and receipts of messages. If the operating system could provide an additional level of indirection in the form of a queue, then a number of other problems such as load balancing, prioritization, and disconnected operation could be taken care of.

The operating system service in the Windows NT environment that facilitates the reliable exchange of messages in a distributed environment is Microsoft Message Queue Server (MSMQ). MSMQ provides a method to locate queues that exchange messages using various messaging guarantees. Security is important in distributed applications and the underlying security infrastructure in MSMQ makes it possible to develop the message infrastructure.

Message Structure

MSMQ mirrors much of the Internet. Internet e-mail uses Domain Name Service (DNS) as a directory service for name lookups. Directory services operate within a defined scope, much as a telephone directory is valid only for a given town or city.

Multiple copies of the MSMQ are stored in different computers throughout the system. In a busy network, the directory service can easily be overwhelmed by lookup requests into multiple sites. Each site has its own copy of the site controller, which provides load-balancing and fail-over schemes.

Messaging

Distributed applications pass messages between computers, either synchronously or asynchronously. With a synchronous exchange, both the sender and the receiver must be running at the same time.

Examples of queues occur in e-mail in-boxes. Messages arrive periodically, and you can respond to each as you see fit. You can view them in this queue.

In MSMQ's recoverable delivery option a message is copied to a disk queue, guaranteeing the message will not be lost. The MSMQ infrastructure is responsible for delivering the message. Transactional throughput is controlled by keeping transactions as brief as possible.

Security

MSMQ supports secure transmission of messages. This involves limiting access to that information. Authentication verifies the identity of the sender. Encryption ensures that the message cannot be read by authorized receivers. MSMQ supports authenticated and encrypted messages.

The two general encryption schemes in use today are symmetric and asymmetric. Symmetric encryption schemes use the same key for encryption and decryption, while asymmetric encryption schemes use different keys.

Both schemes rely on the strength of the key being subject to public scrutiny by the cryptographic community. The main advantage of symmetric encryption schemes is the speed with which clear text can be converted to or from cipher text. Symmetric encryption operates almost 1000 times as fast as asymmetric encryption.

Asymmetric encryption schemes use public/private key pairs. They utilize a defined mathematical relationship between the public and private keys. The main drawback to using asymmetric schemes is the high computational load of the encryption/decryption algorithms.

MSMQ uses a combination of asymmetric and symmetric encryption schemes to encrypt the body of the message. Asymmetric encryption is used to ensure the integrity of the transfer of the symmetric key.

The MSMQ Queue Manager generates a random symmetric key and uses it to encrypt the body of a message. It must also guarantee the authenticity of any public key used. Digital certificates are commonly use to solve this problem. If messages contain sensitive information, they should be both authentic and encrypted to prevent eavesdropping or tampering while they are in transit.

Direct Digital Control

Direct digital control (DDC) technology is constantly changing. In the past several years, there has been a race between controls companies to provide leading edge hardware and software. Many building sites have been converted to or built with direct digital control systems.

Most systems are very close to each other in performance. The latest movement has been towards the ability to easily transfer DDC system information into desktop software packages for data accumulation and processing.

The program is an inherent part of the system. If the computer is not programmed to stringent engineering practices, the DDC system can be a liability rather than an asset. There are many DDC systems in use that have no or little user interface, that function with no comfort problems, and are cost effective. This is because they are quite intelligent. The system was designed specifically for a particular building and functions well under all conditions. Users can make setpoint adjustments without computer interface. With intelligent feedback programming, maintenance personnel are freed from tuning and modifying automation systems on a seasonal basis. Tasks such as maintaining mechanical components, scheduled maintenance, and learning existing building deficiencies can become maintenance priorities. In a truly intelligent program there are no feedforward anticipated schedules, switches, or need for owner intervention to change setpoints and tune loops constantly.

There will be less overall user interface errors that are common when operators must scan programs. There are no points locked in manual since the system no longer requires a user to lock things out or adjust for problem conditions.

Feedforward Errors

Errors that occur from feedforward instead of feedback programming result in higher maintenance costs. Feedforward error conditions include changing summer/winter switches in software and tuning control parameters too often or on a seasonal basis.

Shutting off boilers and then turning them back on when the weather unexpectedly cools down is another one of these problems. Changing outdoor air, boiler and radiation schedules and resetting freeze stats also occur too often.

Modifying programming for cooling unit cycling is a similar problem along with answering cold complaints from cooling cycling due to unstable operation and answering too-warm complaints for single-stage gas valves overshooting. Cooling units can be damaged from incorrect operation and outdoor temperature and boiler loops can fight with cooling loops from feedforward error. Cold complaints in building interiors can occur from shutting down boilers from feedforward programming. Checking software for bugs and the causes for incorrect operations under specific conditions are also common occurrences with feedforward programming.

DDC and Object Technology

DDC moves at a rapid rate and many upgrades are based on the performance results of other field applications. The Visual Icon Design Method (VIDM) is derived from flow charting and desktop publishing.

It uses desktop computer technology to associate control system icons with a programming scheme to form objects. These objects have prewritten, generic code that provide the sequence of control operations. This allows standardized coding and forms an object library. It clarifies the sequence of operations and displays how to achieve it. This provides a framework for standard programming across many vendors as well as better comprehension of the program from various levels. Graphic real-time diagnostics are possible in a graphic environment, using object flowcharts as a diagnostic real-time graphic.

The newer software is engineered for exception entry and requires less work to maintain operations. Audio and video enhancements in the future will make it easy to catalog equipment and describe maintenance procedures. The trend in maintenance management will be towards products that are easily modified.

Open Protocol Issues

Natural forces will define the future of open protocols. The term open protocol refers to a standard communication format based on the exchange of specific modular decisions, which are also called objects, between control systems manufactured by unrelated companies.

Ideally, this could allow open competitive bidding even on large projects that may occur over several years. The concept of open systems originated in the mid-1980s, due to the financial and operational impacts of evolving building automation systems (BAS). Advances in microprocessor technology and the new applications of terminal unit controllers for small HVAC systems, chillers, unit ventilators, and VAV systems, defined rapid advancements and new products that rarely communicated with previous controllers.

Gateways

The need to support existing systems lead to expensive approaches to close the communication gaps with communication gateway devices. These are communication translators and multi-vendor protocol host software systems. They provide central host software functions while communicating with controller networks using dedicated communication ports and software drivers. These gateways and drivers are relatively expensive. All features are provided in the drivers and must be updated when one of the manufacturer's products changes.

Since the manufacturers of building automation systems approached their products and features differently, these proprietary systems had strengths and weaknesses that made them more appropriate and competitive for specific applications.

End-users preferred the products they had experienced, because they have been trained to make these products work. The costs of training for multiple systems, spare parts maintenance, along with employee turnover, cause multiple-system applications to be less desirable.

Users desire a standard method that ensures unrelated products can communicate with each other under a competitive bid structure. One problem is that systems already installed will not easily or completely become compatible without gateways, software drivers, or replacement. Buildings often have existing computer networks that can be used as a communication medium for the building automation system.

Protocols

The Building Automation Controls Network (BACnet) protocol was developed by ASHRAE. BACnet is a multi-layer Open System Interconnection (OSI) basic reference model. It is a set of rules that allows unrelated manufacturers to communicate in predefined data packets. The GSA has made BACnet a requirement in their buildings. Early concerns about large data packet transfers under BACnet have been proven unwarranted.

Another protocol is the Local Operators Network (LonTalk) which was developed by Echelon and implemented on an integrated circuit called the neuron chip. In the chip are three microprocessors. Two are used for networking and one for application-specific tasks in control devices.

Open Control Devices

Thermocouples are one type of control device. Biodata's TC15-Lon is designed to monitor temperatures. Providing 15 channels, it connects to a LonWorks network using 4-wire telephone cable which carries both low voltage DC power and data. The unit reads temperatures and checks whether they are changed since the last reading. It passes any new data to the PC, with instant notification of alarm conditions. Since it only transmits changed data, the network traffic is minimized. You can move the device around without worrying about where it is on the network. It always identifies itself.

The device comprises an intelligent isothermal box. It has 30 screw terminals for connecting the thermocouple wires, a sensor to measure the temperature of the unit (the cold junction sensor), an A-D converter and a processor. It can handle B, E, J, K, N, R, S and T type thermocouples.

A Windows software package called Windmill is included with the system. It provides data logging, charting, output control and real-time links to other software. A distributed thermocouple system can increase the accuracy of temperature measurements. Resistance temperature devices (RTDs) are also available as Lon units.

Open Building Controls

As standards of open interoperability continue to evolve, some building control systems promise compliance with several standards. Siebe Environmental Controls has their I/A Series with flexibility and scalability in LonMark and LonWorks controllers. WorkPlace is used as the Human Machine Interface (HMI). Based on Wonderware, WorkPlace provides single seat control of building systems. The system is optimized and configured using an integrating driver with the Tech Suite or by adding capabilities through Microsoft Active XCOM object technologies.

The Lon Network Controller provides connectivity, alarming, time of day schedule optimization, maintenance reminders and global information management. The Tech Suite module uses graphical layout and presentation technology, combined with the intuitive assistance of VISIO. The MicroNet controller family allows tailoring of the control strategy.

As technologies move forward, they should remain focused on backward compatibility to assure that users have an opportunity to take advantage of their installed infrastructure. Connectivity ranging from dial-up PBX, to full integration at the workstation and control system levels are expected by today's building owner. The I/A Series addresses these requirements and provides client/server functionality through

Windows and NT. With dial-in and auto-dial telecommunication twisted pair or thinnet connections can be used.

Windows Control Software

Wonderware uses Microsoft Windows for their human-machine interface (HMI) software. Wonderware software includes InTouch as a (HMI) for visualization, InControl for Windows NT-based control and Scout as an Internet/Intranet tool for remote data viewing.

InTouch provides a single integrated view of all control and information resources. It allows operators to view and interact with the workings of an entire operation using graphical representations. It includes remote tag referencing, ActiveX support, distributed alarm handling, and distributed historical data. Individual alarms help users to quickly identify rules that were violated. Users can track corrective actions and set limits on the fly. A high-speed communications protocol, based on TCP/IP, is used to provide performance-optimized data communication in the Windows NT environment for real-time, peer-to-peer communications.

InControl is an NT-based real-time open architecture control system that allows the user to design, test and run application programs for controlling a process. Direct interfaces to a variety of I/O devices, motors, sensors, and other equipment are available.

The Internet visualization module acts as a web server add-on and client browser that allows read-only remote viewing of data and visual objects over the Internet/Intranet. In the client or browser mode, visual objects can be viewed from the server using a standard browser. The software extends a browser's capability by providing a set of user-configurable ActiveX objects such as graphs, charts, and trend elements that can be dynamically linked to the data. In the server mode a set of web server components runs on Microsoft's Internet Information Server (IIS) and links applications to the Internet.

Building managers may soon see their roles expanding with the responsibilities for all electrical systems. Just as the division between telephone and information systems has largely disappeared, process control and facilities management are also headed for closer links to computer networks. This convergence between applications is made possible by open systems that can control manufacturing equipment, elevators, heating and cooling systems, generators, security and fire systems, pumps, parking lot gates, and other electrical systems through intelligent devices connected to that equipment.

References

Bernardi, and Brent S. and Dan R. Epperson, "I/A Series, Enterprise-wide Open, Interoperable Solutions, Today and Beyond," *School Planning and Management*, Vol. 38, No. 1, pp. 22-24.

Cleaveland, Peter, "Motion and Position Control in the 21st Century," *Instrumentation & Control Systems*, Vol. 73 No. 1, January 2000, pp. 21-28.

Glinke, Timothy J., "The Open Protocol Choice: a Market's Decision in Process," *Air Conditioning, Heating & Refrigeration News*, Vol. 202 No. 6, October 6, 1997, pp. 6-9.

Gloriso, Robert, "Tolerant to a Fault," *Industrial Computing*, Vol. 18 No. 11, October 1999, pp. 29-33.

Hawman, Chris, "Intelligent Alarm Management Saves Industry $$$," *InTech*, Vol. 44 No. 10, October 1997, pp. 34-36.

Lam, John, "Improving the Performance of Distributed Applications," *PC Magazine*, Vol. 18 No. 3, pp. 247-253.

Liptak, Bela A., Editor-in-chief, *Instrument Engineers' Handbook*, Revised Edition, Chilton Book Company: Radnor, PA, 1982.

Liptak, Bela A., Editor-in-chief, *Instrument Engineers' Handbook*, 3rd Edition, Chilton Book Company: Radnor, PA, 1995.

Oshoway, David, "An Overview of Direct Digital Technology," *Air Conditioning, Heating & Refrigeration News*, Vol. 202 No 7, October 13, 1997, pp. 14-17.

Phillips, Ken, "First Telephones, Now Lights," *PC Week*, Vol. 16 No. 17, April 26, 1999, pp. 77-78.

Risse, Lori and Greg Vonada, "High Accuracy Temperature Measurements: Are They Worth the Investment?," *Instrumentation and Control Systems*, Vol. 72 No. 6, June 1999, pp. 31-34.

Thompson, William E., and Richard H. Caro, "IT Revolution Reaches Industrial Automation," *Control Platforms* (Special Supplement to *Industrial Computing*), October 1998, pp. 24-27.

Internet: www.microlink.co.uk/tc15lon.html, "Distributed Temperature Measurement Using Thermocouples or RTDs and LonWorks Network," February 22, 2000, pp. 1-2.

Chapter 5

PC-based Control, Software and Bus Trends

PC-based control is becoming a mainstream technology, yet there is still some controversy between programmable logic controller (PLC) manufacturers and PC-based control suppliers on the use of PC-based control technology. PC-based control is much more than simply emulating a PLC on a PC. PC-based control systems using only traditional PLC programming languages realize only part of the technology's real benefits. These benefits include the integration of the control system into a single tag database.

This integration combines the functions of logic control, programming, operator interfaces, supervisory data collection and enterprise interface in a single platform. A common tag database among each of the functions eliminates multiple databases and the data communication configuration needed for multiple components. Another major benefit is the ability to use off-the-shelf PC hardware and software. PC-based control offers a lower-cost control solution and continuous efficiency improvements.

Data connectivity between applications on the same or different platforms is possible through interfaces such as OLE for control (OPC), distributed communications and dynamic data exchange. OLE stands for object linking and embedding.

Object Concepts

An object is a software concept that often embodies hardware elements. Objects are a more natural way of solving problems. Humans observe and interrelate on an object basis.

An object knows its purpose. It responds to stimuli (inputs), performs its function, which may be algorithms or methods, and sends messages to other objects as its outputs.

Objects are event-driven, waiting for a change to start an action. They do not know anything about the objects that provide the inputs or the objects that receive their outputs.

They may be replicated to extend functionality and replaced to improve performance or change behavior. Removal is simple when the functionality is no longer needed. All of this can be done without impact on the overall project function.

Objects are tolerant since they do not care about the makeup or complexity of other objects. A control object could be a PLC, since the PLC can be encapsulated and treated as an object.

Object based solutions can be deployed alongside existing traditional solutions and when it is desired to replace existing systems with new ones, it can be done smoothly and effectively.

Objects are trusting since they always assume that the downstream and upstream objects will always do their tasks. Objects are widely distributable since they do not require other objects to reside on the same computer platform and they do not mind if other objects are relocated.

Objects behave the same if they function on their own or are a part of a complex control solution. They scale up to meet the demands of the situation.

Object-based solutions have been used to automate many plants and process solutions in the oil and gas industries. In these, complex sequential operations were used to ensure consistency and completion. The object-based solutions that replaced them were easier to implement and maintain and perform more effectively than the systems they replaced.

Operating Systems

In this information age, personal computers have become an everyday part of life, both at home and at work. But, these computers are useless unless you can easily access the information.

Operating systems provide an interface between the user and the computer, allowing the user to access and control the computer resources. The operating system controls the access to the processor and interfaces with the different system resources such as hard and floppy disks, CD-ROMs, printers and modems. The operating system acts like a traffic cop between the hardware and the application software programs. The operating system controls basic tasks such as directing data to and from devices.

As operating systems have evolved, they have taken over functions that the older applications software used to handle. During the last few decades, the quality of operating systems available for small computers have increased with the demand for better user interfaces. One of the most popular operating systems for control systems is Windows NT.

PLCs and PCs

When PLCs were introduced over 30 years ago, relay ladder logic (RLL) was the standard programming language. Ladder diagrams identify the order of operation of devices. The logic can be rewired by programming on the CRT display and tested, without any physical rewiring of the banks of relays. The displays show a ladder diagram with contacts, coils and circuit connection very similar to a schematic diagram. RLL was not designed to support a wide range of control functions.

The main input devices to PLCs are push-buttons, limit switches and other command devices. The output devices are contactors, relays, solenoids and indicating lamps.

PLCs have been moving towards commodity hardware items, as PCs already have, and the trend is towards providing more complete application solutions. This means more open systems and commercial off-the-shelf technologies to meet user demands for performance, ease of use and lower cost of ownership.

Continued optimization of overall plant performance requires leveraging plant information with automation applications. Even PLC architectures are increasingly being dominated by the flow of information. The use of object-based technologies, industrial Ethernet and Internet access to information are examples of this trend. PLCs along with PCs are changing from closed-control-specific devices to open information servers providing access to previously untapped operational information.

Flowchart Programming

Flowchart programming is an advance that helps to realize the full power of PC-based control systems. A simple, intuitive graphical description of the system is provided. Predefined flow elements and a graphics builder are used to create custom objects specific to the application.

Operator-level diagnostics have been difficult and time consuming to setup with ladder logic. The diagnostic tools available in flowchart programming can provide diagnostics that automatically generate fault detection and recovery logic, along with displayed operator error messages and HTML-based operator repair and recovery displays. These tools are important for improved efficiency since they provide quick recovery instructions to operators which substantially lowers equipment downtime.

Integrated motion control is another benefit of flowchart programming. Motion control is one of the most significant features to be integrated with logic into PC-based control systems. You no longer need to configure a separate motion controller with complex programming. It can be done inside a PC-based control system with a single dialog box.

Embedded Devices

Many advances in computing power are taking place in portable and embedded devices like controllers and intelligent sensors. Unlike desktop computers, these smaller units have little use for the power and size of an operating system (OS) such as Microsoft Windows 95/98 or NT. They need an OS that is easily adaptable, compatible with a variety of processors, and can operate without a terminal or keyboard.

Microsoft's Windows CE was developed to fill these needs. In addition to being operable without a terminal or keyboard, this reduced version of the Windows OS provides support for more than 20 processors from 11 manufacturers and it uses Win32 application programming interfaces (APIs).

Win32 Programming

Developers of applications for Windows can use Win32 programming resources. The Win32 APL defines the interfaces to the Windows platforms that are available to a programmer.

Programming using Win32 interfaces is not the only way to create

32-bit, Windows-based applications. Another programming interface available to Win32 and Microsoft Visual C++ programmers is the Microsoft Foundation Class library (MFC). MFC provides higher-level encapsulations for much of the Win32 API.

MFC supplies C++ classes to represent Windows user interface objects such as windows, dialog boxes, brushes, pens, and fonts. MFC class member functions make calls to the Win32 API functions. The programmer using Win32 can program using C or C++ and the Win32 API, or using C++ and MFC.

Windows CE

Windows CE is not just a reduced-feature-set version of Windows. It was developed as a small-footprint OS for embedded applications. Its kernel borrows from other Microsoft 32-bit OSs and eliminates or replaces those OS characteristics that typical CE-based applications do not require. See Figure 5-1.

Windows CE applications, like Windows NT applications, run in a

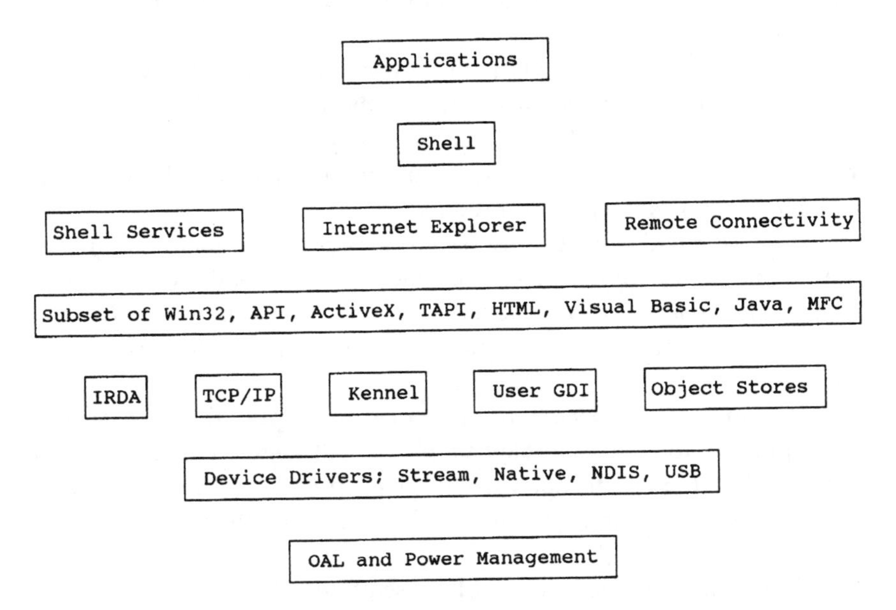

Figure 5-1. Windows CE architecture with an OEM abstraction layer (OAL) for device-dependent code and device power management.

preemptive, multitasking environment, in protected memory spaces. CE is compact and can be contained in less than 200K of ROM. Windows CE's Win32 API has about half of the interface methods of the Windows NT version of the API.

The CE API allows application notification events (timer events) at the OS level, rather than in a running application. CE supports touch-sensitive displays. The Windows graphic interface in CE makes it particularly useful in devices that need a standard Windows interface.

Windows CE and NT address different applications. CE offers a simpler multitasking model than NT while NT offers more powerful communications functions. Windows CE was designed as a modular OS, so it can be adapted to many different applications such as control, monitoring and data collection. Built-in communication features include: secure Internet, wireless communication interfaces and extendible device capabilities (Figure 5-2).

Windows CE is a useful platform for operator interface and control. Mobile data collection device manufacturers are using Windows-CE-based products. Windows CE is also well-suited for open PLC/controllers, intelligent field devices and instrumentation, utility monitors, VFDs, smart motor controllers, quality auditors and building maintenance.

Much of the functionality in the new hand-held devices will be due to the human-interface characteristics of the universal Windows environment. This functionality will be migrating to other applications as Windows CE finds its way into more and more embedded designs such as Web TV and home-control products.

Low cost and functionality have been among the main features of the PC and portable designs have followed that trend. As industries require more users to be mobile and use portable gear more, the need to provide more employees with portable units will help to drive down costs.

The increased expectations of users as they apply these productivity-improving tools to everyday problems will demand an increase in functionality. Users will not want to give up much of the power and diversity of the desktop as seen by the huge growth in notebook-PC capabilities.

The trend is moving toward the use of knowledge workers. These are employees that are expected to manage information from a variety of sources, turn that information into usable knowledge, and apply that

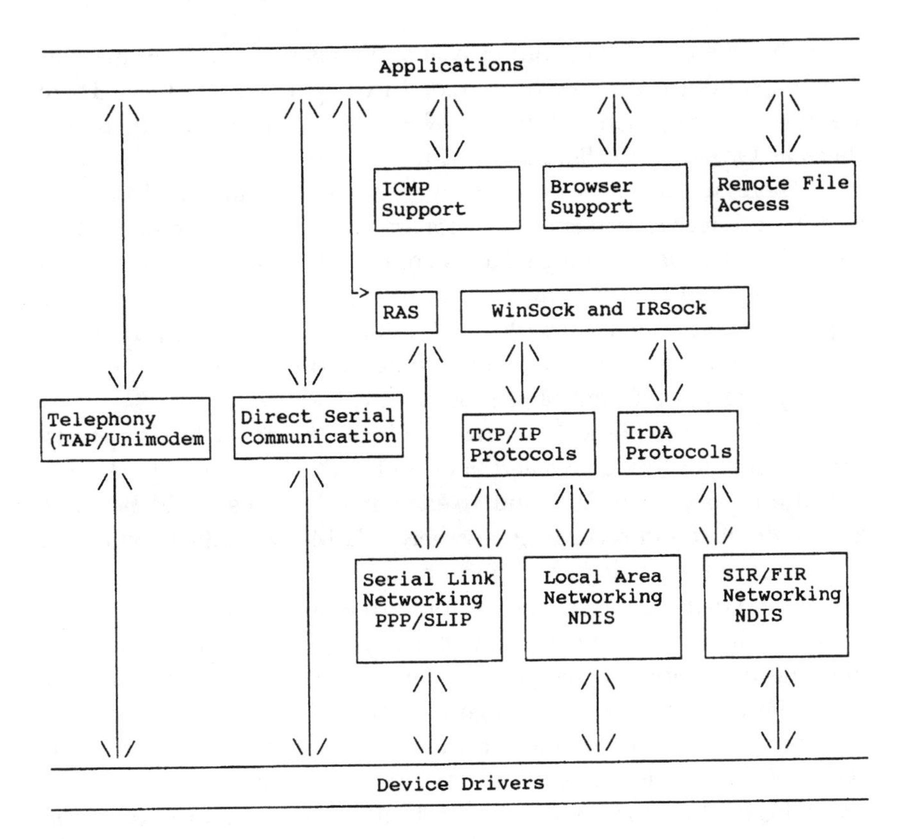

Figure 5-2. Windows CE supports communications protocols for telephony and serial communications, as well as TCP/IP and other Internet protocols.

knowledge to the solution of the problems.

The enabling technology includes interconnectivity of these handheld units with other PCs, the Internet and the company Intranet. All of this functionality is expected to be seamless for the user. A familiar Windows or Windows-like graphical user interface (GUI) will be required. Interconnectivity with desktop and laptop PC products will become a growing requirement.

The impact that Windows CE will have on portable designs involves several emerging areas. These emerging developments are in microprocessor unit (MPU)/multipoint conferencing unit (MCU) architecture, new human interface options, greater connectivity with Internet and PC products and the growing refinements in programming.

The Windows CE operating system (OS) is modular, so some portions of Windows CE may be excluded from portable products. There are power-management schemes in Windows CE that define the power states as On, Idle, and Suspended/Off.

There are also new ways to interact with portable and desktop PCs. Speech recognition and synthesis are used for heads up interaction and safety. The personal digital assistant (PDA) of the future is likely to have some kind of handwriting recognition, similar to the input made popular by the 3Com PalmPilot. Full-motion video will become a requirement for Web use. Video input could be used for face-recognition security and gesture control of programs.

PDAs may be used to change the heating and cooling setpoints, turn on lighting, appliances and boot up the PC. Hand-held PDAs will communicate with mobile and fixed networks. This could be done across a robust and secure wireless network. Many products will need more wireless technology built into them.

The advent of Windows CE forces the additional integration of systems. There will be a merging of pager, phone, voice mail and e-mail into one unit. More people will be able to use these products as they become more user friendly through the familiar Windows environment.

Performance will become critical. The proper configuration of the system (power management, cache size, memory speeds, bus latency, and CPU architecture) will allow 100-MHz products to outperform poorly-optimized 233-MHz systems. Improved interconnectivity will allow the user to be able to store files, web pages, and messages on the server instead of the PDA.

There will be continued improvement in storage medium for the PDA. Information overload will continue to grow, and information processing is driving technology. PDAs will be enlisted to control this overload.

A typical PDA device will use two major integrated-circuit (IC) parts; a Windows CE million-instructions-per-second (MIPS) processor and an analog application-specific integrated circuit (ASIC) for input/output interfacing.

Windows CE is language-independent and compatible with more than 2,500 of the most popular Windows APIs and Win32 programming models. It uses the same .exe/.dll files, processes, and threads. It can be stored in ROM and has built-in communications to PCs, the Internet, and other Windows CE devices.

Windows CE uses a graphics device interface with more than 50 available components. The use and arrangement of these components allows application interfaces for specific products.

No longer being tied to the X86 microprocessors is a new development for Windows products. Windows CE has been ported to at least five processor families including MIPS, SH3/4, PowerPC, and Arm/StrongArm processors as well as a variety of peripheral devices, including PCMCIA, PC Card, PCI, and ISA buses. I/O types include modems, Ethernet, serial, parallel, infrared, keyboards, displays, flash memory, rotating media, wireless, touch screens and voice. Device drivers are used to support custom peripherals. There is also device power management.

Windows CE provides graphics, windowing, event management, support for color, fonts, dialog boxes, shell, printing, sound, I/O, and files. Windows CE can run 32 protected processes and an unlimited number of threads. With one processor, it supports preemptive, priority-based scheduling, and has eight priority levels with the ability to handle priority inversion.

CE uses virtual memory where memory is allocated to applications one page at a time. The page size is determined by the application. In a hand-held application, the memory page size is usually 1K-4K bytes.

Many Windows 95-based applications can be ported to Windows CE. Only a subset of Win32 API is supported and some of this has a reduced feature set with fewer supported window styles and limited support for colors and fonts.

Development

Windows CE development tools for system design include Platform Builder with OS modules and sample configurations. It is set up for PC-based or Hitachi D9000 reference platforms. The Platform Builder supports object-oriented programming through Microsoft's Component Object Model (COM). Other development tools include sample OALs, device drivers and a configuration tool to select only the components that are needed for an application. A ROM image maker and remote debugger are also provided. The tools provide the ability to develop applications using an emulation of a real device.

Language support for application development consists of the same language-based tools used for standard Windows 9x/NT programming. These include Visual C++ and Visual Basic.

Processes and Threads

In Win32 terminology, a process is defined as an instance of a running application. Windows CE, like other 32-bit Windows platforms, is a multitasking OS that supports multiple threads of execution within a running process. Thread management includes scheduling, synchronization and resource allocation. These are handled by the kernel.

Inside the processes that run on 32-bit Windows platforms are the threads that rely on messages to start processing, control system resources and communicate with the OS and user. Windows messages can originate from a variety of sources, including the OS, keyboard, mouse or touch screen actions and other running processes or threads.

Exception and Device Handling

Sections of code that could fail due to hardware resource problems, device conflicts or coding errors need to be protected. Exception handling protects the application from termination or other system-level problems.

The software interface must be able to manage simultaneous or nearly simultaneous events from different types of devices in the system. The Win32 API supports both synchronous and asynchronous methods of device access. In a synchronous interface, the software requests an action from a device and then waits for the result. Examples include reading and writing files. In an asynchronous interface, the device requests services from the application. An example of an asynchronous device is a keyboard.

In the Windows CE kernel, interrupt handling is split into interrupt service routing (ISR) and the interrupt service thread (IST). Keeping the routing as small and fast as possible is important for real-time applications.

At the hardware level, each interrupt request (IRQ) line is associated with one specific software ISR. When triggered, the ISR directs the kernel to the location of the IST.

Interrupts have a priority associated with them. Windows CE uses a priority-based, time-slice algorithm for thread scheduling. The Embedded Toolkit allows tuning of the interrupt timing and priorities for a specific application.

Windows NT-2000

Soft-logic, PC-based control can run on NT systems without the protection of a real-time kernel. The kernel is the heart of the operating system. However, using a real-time kernel assures deterministic performance and the control system is also protected against NT failures, buggy Windows drivers and hard drive crashes. If Windows NT is used with PC-based control, it provides a familiar interface for the operator. The real-time aspect is separated from NT, so if you lose NT, you lose the interface but the control can still operate. The real-time engine is hidden from you and does not impact any programs running on NT.

Interchangeable I/O drivers are a feature of most open PC-based control systems. Users should be able to switch I/O families and networks as easily as changing printers in Windows. A dialog box allows older I/O systems to be changed to more modular, lower-cost I/O. An I/O rack is used for the various input/output modules. I/O modules are available is many configurations including analog, communications and motion control.

Windows 2000 was formerly Windows NT 5.0. Changes include simplified application utilities that make it easier to use NT. It is now more similar to the standard Windows interface. Microsoft has also made some changes which should improve future OS system upgrades. These may be a little troublesome now, but ultimately they will improve the reliability and integrity of the operating system.

A major change is the Windows Driver model (WDM). This change should allow more flexibility and enhance reliability. VenturCom has a special driver model for automation and control with real-time capability. MicroSoft's driver includes plug-n-play which makes it easy to add cards for PCI-based communications and other I/O.

DirectX

Part of the WDM is Microsoft's DirectX multimedia architecture (Figure 5-3). This architecture works well in Windows 95/98, but the use of it in Windows 2000 may affect the stability of the OS. DirectX offers a fast interface to hardware since it is really a driver model. These drivers could be virus portals and should be replaced in critical industrial projects.

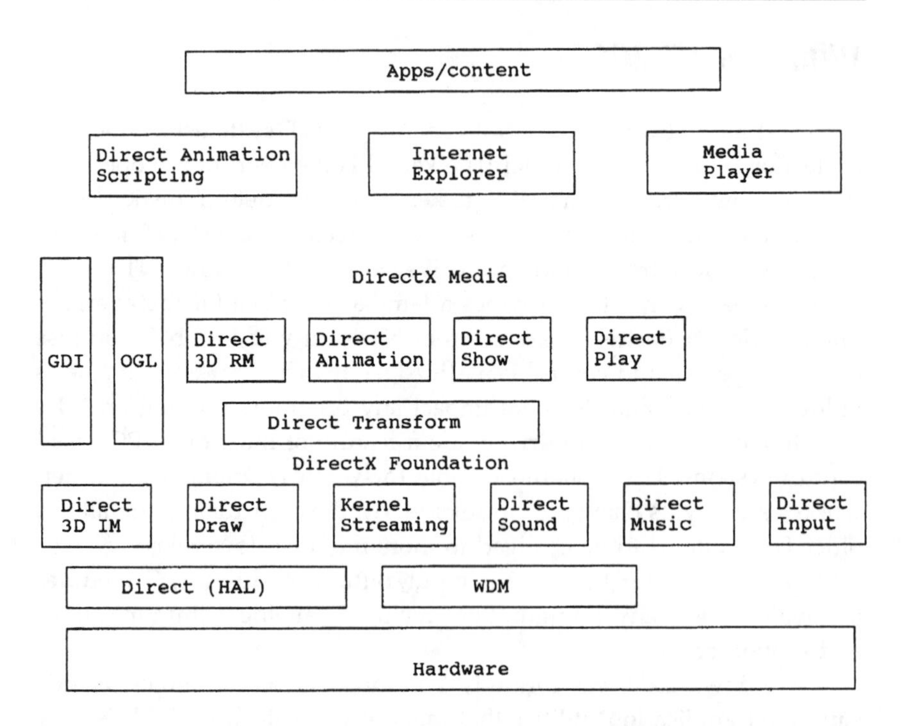

Figure 5-3. Microsoft's DirectX multimedia architecture provides everything needed for a good operator interface.

Microsoft runs video drivers in the kernel of Windows 2000 and NT 4.0. In Windows 2000, this helps high-end graphics applications, such as operator interfaces, but may cause problems in controlling a process on Windows NT. Standard VGA drivers can be used in critical NT applications.

Microsoft has not been able to provide all of the DirectX capabilities on NT due to security concerns. The driver model needs to protect all the interfaces to the OS. If the interface to the hard disk is not protected properly, loss of data and the system can occur. Windows 2000 has significantly improved real-time performance capabilities over Windows NT 4.0.

Windows NT and Windows CE both have their place in control. They can provide complementary control solutions. Think & Do Software has a PC control software package for control and (HMI) that is scalable for both Windows NT and/or Windows CE. This scalable approach allows both NT and CE platforms to use the same code in small stand-alone machines or fully automated process networks.

Another approach combines Windows NT with real-time extensions. VMIC has IOWorks which is a PC-based control software development package for these real-time extensions.

Windows NT can also be used with a real-time operating system like VXWorks. In the case of VxWorks, two computers are involved; one is used for Windows NT development and HMI and a VxWorks computer handles the real-time application. This provides an acceptable PC-based development environment with the security and performance of a proven real-time operating system like VxWorks.

CSI has a User Configurable Open System (UCOS) that was designed to run QNX on PC-based hardware. The QNX OS is mature and reliable and easy to integrate with Windows NT-based engineering and operator workstations.

UNIX

Before Windows, UNIX was one of the dominant operating systems. In this era of open systems, some vendors such as ABB, CSI, Foxboro, GE Fanuc, Honeywell, MDT, USDATA and others still offer UNIX products. GE Fanuc's CIMPLICITY is based on VMS and UNIX. CIMPLICITY was rewritten for the Windows NT platform and released in 1995.

UNIX retains a considerable legacy base. But, while UNIX systems can provide a role in database support, UNIX systems lack the available integration technologies such as those found in Microsoft products.

Some software vendors that do not directly support UNIX applications see the need to communicate with them. The newer methods for connectivity are mostly in the form of Microsoft provided connectivity for databases on non-Microsoft platforms. These are typically UNIX-based databases.

Linux

Linux has been growing as it moves into more and more applications. Many of these are through the back door as an inexpensive, high-performance, unbreakable server platform, which is also capable of multiprocessor and cluster implementations. It also exists in a real-time version and has earned a reputation as a mission-critical server platform (Figure 5-4).

Several vendors are prepared to support Linux. GE Fanuc CIMPLICITY has its UNIX experience and may support Linux as well. National Instrument's LabVIEW is on Linux and Opto 22 support Linux with Linux drivers for its Ethernet and PAMUX I/O products.

The Linux operating system has to reach commercial levels of success for desktop systems in order to be fully accepted. Many server products use Linux and the level of reliability and robustness needed is one of the toughest.

Real-time Linux

Real-time Linux uses a real-time executive, which takes priority over the main Linux kernel. The purpose of RT-Linux is to mix two

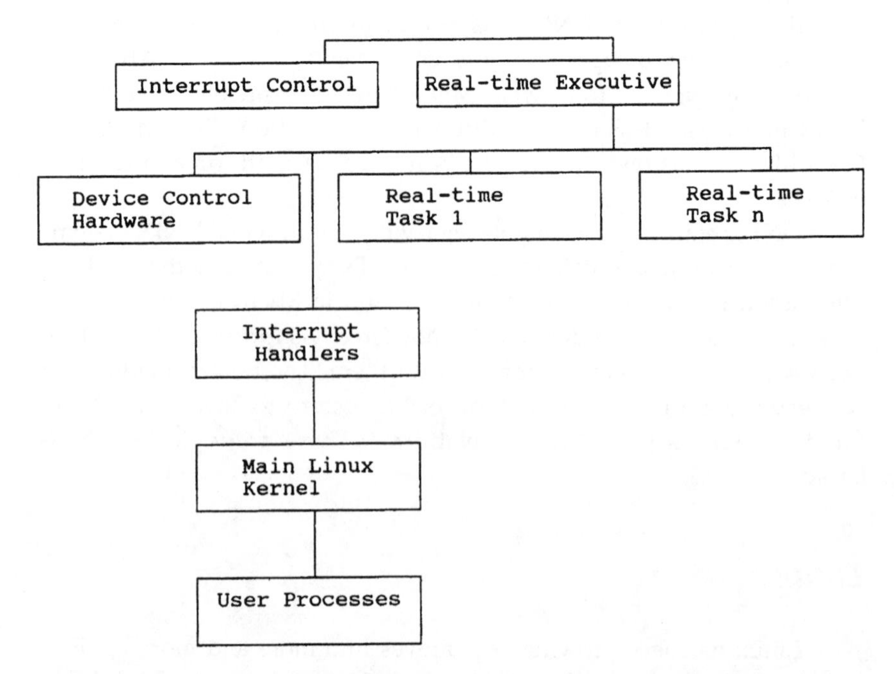

Figure 5-4. RT-Linux with a real-time executive that runs the main Linux kernel as a subordinate task.

incompatible properties in the same operating system. One is hard real-time service with its predictable, fast, low latency, simple scheduler. The other is the services of GUI, TCP/IP and web-servers.

Linux is a low-cost OS with a high-performance microkernel architecture. Linux is the ultimate example of open architecture, with hundreds of thousands of end-user contributions.

Object Technology

The trend towards open source software continues with object technologies and Internet protocols making it possible. There were no standards available for object technology back in the 1980s, so the real-time Object Manager was developed. This type of object-based software and communications infrastructure provides a high degree of hardware and operating system independence.

Today, there are many industry and defacto object standards such as Java, CORBA, OPC, Active X and DNA. Each has its place within the complete sensor-to-database automation solution. There has been much success in using Foxboro's Object Manager with these industry and defacto object standards to create object-enabled environments.

Many of these applications can run on a number of operating systems including Windows NT, SUN OS, Solaris, VRTX, and Venix. Standard TCP/IP networks, OpenVMS, HP-UX and AIX are also supported.

Several object-oriented technologies are playing important roles in control software. These are OPC, ActiveX and Java. The latter two are finding their roles primarily in networks and the Internet. CORBA and ORB are also object technologies, but these are found primarily on UNIX systems and typically work at enterprise levels. They have not seen major use in industrial data communications.

OLE (Object Linking and Embedding) was Microsoft's original object-based communications technology within Windows. OLE allows you to assign tasks to objects and create integrated applications that can share data.

OLE was replaced by the newer OPC (OLE for process control). In process control, OLE/OPC opened up communications between devices and software applications in the Windows environment. OPC became the standard technique for exchanging data between clients and servers. OPC toolkits from companies such as PC Soft extended the technology to Visual Basic and other languages.

OPC has become the main technology for connecting industrial software applications. OPC makes small custom applications easy to generate and support (Figure 5-5).

Proprietary object technologies have been used by companies such as CSI, Foxboro and Honeywell. OPC is now used extensively throughout Honeywell's products.

It provides a standardized function, which allows many new kinds of information integration to occur although it is not as sophisticated as many existing proprietary access schemes. The availability of a more basic and functionally inferior standard access mechanism has been enough to replace many older proprietary schemes. One problem with

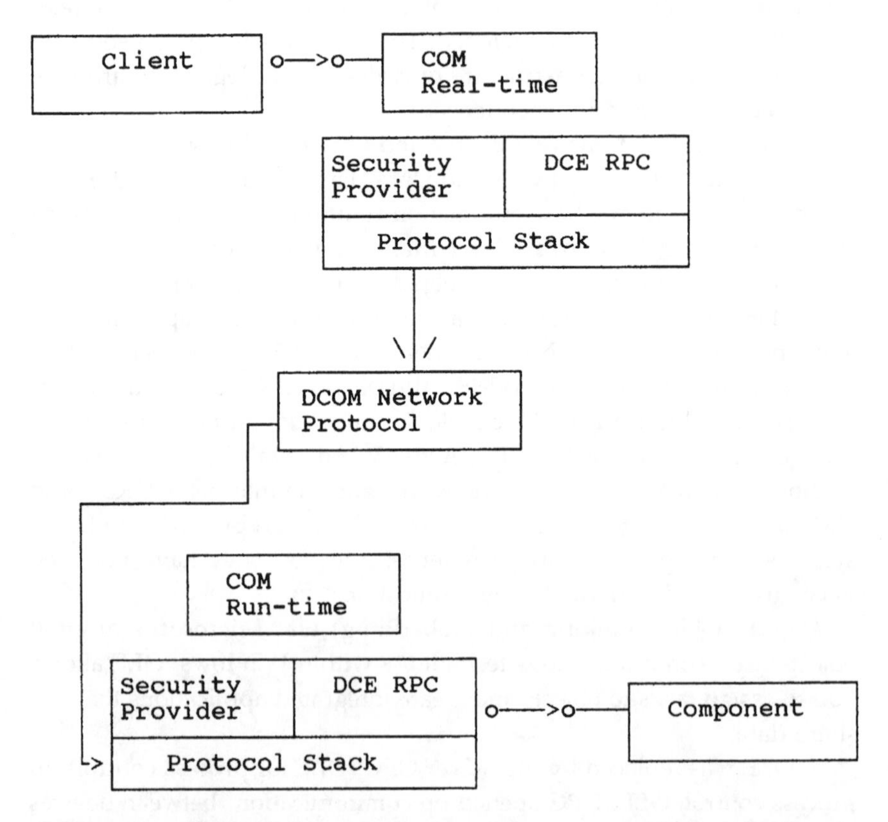

Figure 5-5. OPC has its roots in the Microsoft DCOM (Distributed Component Object Model) technology, which communicates over networks, and is part of Microsoft DNA.

OPC has been speed. In some control software OPC could be too slow to use.

OPC may become the generic data exchange mechanism for industrial automation. For zone automation, it is not the platform that is most important, it's communication. Where performance is needed in the PC-based control, DCX enhancements can be used to supplement OPC.

DCX

DCX is a newer architecture that may be adopted as a standard real-time communications specification. OPC and Microsoft's COM (component object model) can provide enterprise support for data communications, but do not provide real-time deterministic response.

DCX uses a publish-and-subscribe based model to provide driver shells for common devices, hard real-time development tools, standard configurators and OPC. DCX offers CE and NT portability, provides OS isolation, offers plug-and-play with applications and drivers from different vendors and supports OPC connectivity (Figure 5-6).

Component Software

Component software is a major concept for new application development. The two basic models are Microsoft's component object model (COM) and the Object Management Group's common object request broker architecture (CORBA). The Object Management Group is a consortium of major software and hardware vendors.

COM is used to build systems based on the Microsoft Windows environment. CORBA is used to build systems using more diverse operating environments such as IBM's MVS, CICS, Unix versions or Windows NT.

Microsoft has been developing the technologies for building components in the versions of OLE, COM and ActiveX. These standards must be implemented in either Microsoft Foundation Class or Active Template Libraries.

Component developers have focused on developing user interface components. These include component libraries of charts and panels. Developers of industrial automation software have extended that con-

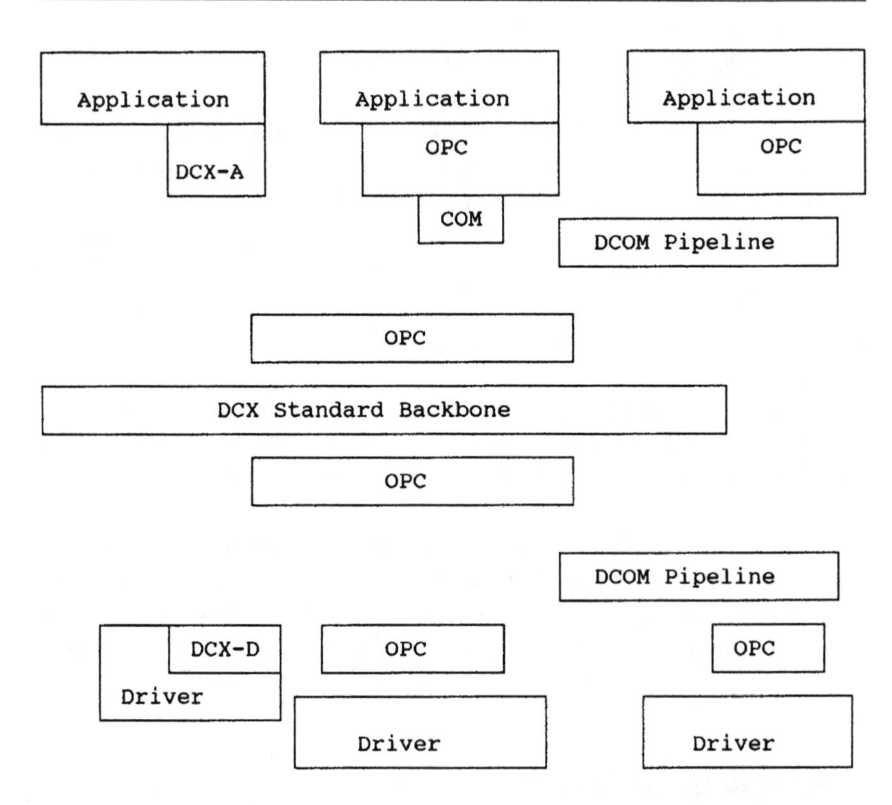

Figure 5-6. VenturCom's DCX is a new architecture that can provide fast, deterministic data transfers under OPC.

cept by developing component libraries that model gauges, dials and thermometers.

This is just a taste of what is to come. There will be much more of these software components and component libraries that model controls and control systems.

User interfaces will resemble process dashboards using a library of components that model pumps, proportional-integral-derivative (PID) loops, tanks and other system blocks. Systems integrators who have an expertise in a particular industry can capture that expertise in a component software product.

There is a trend to develop systems in an iterative way that reuses components and maintains flexibility throughout the system's life cycle. This is called rapid application development. It depends on developing iteratively to reduce risks by exposing them early and emphasizes the importance of software architecture and software reuse. Object-oriented methods are used in day-to-day analysis, design, implementation, and maintenance tasks.

There is a major technology shift from waterfall-style development and structured methods to iterative development and object-oriented methods. In waterfall-style development, the analysis, design, coding, unit-testing, software-integration, and delivery phases occur sequentially and the software is delivered at the end of the process.

Iterative development is more economical and predictable since you specify the objectives for the software system and then build and deliver a series of partial but increasingly complete implementations. These implementations, or iterations, are working products en route to the completion of the tuned project. Using iterative development, you integrate the software at each iteration instead of at the end of the project. This frequent integration reduces risk by exposing it early in the project life cycle.

This concept, which is used by most major vendors like Microsoft, allows feedback from users who exercise the software delivered at each iteration. Users can actively participate in the refinement of the software through all phases of the acceptance process.

Each iteration exposes problems and risk in the system including functionality, performance and user interface. The remaining elements of risk become the focus of the next iteration.

ActiveX

ActiveX is based on Microsoft's COM. It can be used for connecting different vendors' applications and serves as the glue that connects an application and makes it versatile. One example is USDATA's systems where they are built upon a COM/ActiveX foundation. Each application is a collection of components that is part of the system. This technology is used with an object model to allow a flexible and open environment.

Northwest Analytical's software also uses ActiveX objects. Data may come from a variety of applications, but it can be presented in the same standard way regardless of source application. The use of ActiveX provides the necessary components for this capability.

The use of ActiveX brings up the question of security, since an

ActiveX component is actually a program that can access a Windows-based computer with the same privileges as any program. Honeywell uses several security measures:

- digital signing of components gives assurance to users of the authenticity of the controls,
- integrated NT security provides user authentication through the system and
- the kerberos 5-based security available in Windows 2000.

ActiveX and Java

ActiveX is still proprietary to Microsoft, so to connect to other systems, there may be the need to support both ActiveX and Java. Cube Technology supports Microsoft technologies and their evolution toward ActiveX and COM protocols in products like the CUBE-WEB module which has the multiplatform benefits of Java.

PC Soft also uses Java for its Internet-enabled applications. This provides the ability to run on many different operating systems which is a limitation of ActiveX when accessed by a browser.

Fisher-Rosemount has built its DeltaV control system primarily with OPC, but Java is used in its DeltaV Web Server to provide Intranet connectivity. Fisher-Rosemount also has a DeltaV Web Server tool which allows users to convert DeltaV operator graphics and trends into read-only Web pages. This permits access to DeltaV operating information using standard Internet browsers on PC hardware. This reduces costs as well as the difficulty of accessing the data.

Java may be continued to be directed by Sun in desktop applications, but the direction and utility in embedded applications will probably be fractured. A number of companies now offer good Java Virtual Machine implementations.

Internet: Key to the Future

The Internet and its related technologies will be much more important in the future. Sequencia uses Web-enabled documents where operators can use standard Internet browsers to monitor and record operations. The technology also uses hand-held wireless terminals.

Web integration is becoming more important because of the many advantages offered by this approach. By using a Web browser, additional software does not need be installed, since users are already familiar with these interfaces. Full integration with existing Internet/Intranet resources is also easy to accomplish using this approach.

Security for Internet-enabled I/O devices includes firewalls and routers. When the I/O device needs to be stand-alone, it is possible to use hardware/IP address filtering to limit access to known users or to use encryption schemes like the Web-standard Secure Sockets Layer. The type and extent of security depends on the device resources and the application and/or sensitivity of the data.

Maintenance

One issue to consider is the cost of maintaining a PC. This cost has continuously increased while the cost of the PC has decreased. The cost of maintenance has increased as the complexity of the software has increased.

When software is installed on a PC, there is a moderate chance that either it will not work, or an existing application will stop working due to hardware conflicts, software conflicts, software bugs in the install program, or operator errors during the installation.

Microsoft provides scaled approaches to this problem that range from hand-held devices to low-power PCs to high-end Windows 2000 workstations and servers. Other operating systems such as Linus provide a scaled, build-to-need approach, that generally requires less hardware than Windows-based systems.

Web-based applications built on Java and/or ActiveX will require fewer resources, making systems less expensive and more reliable.

Future Computers

Computers will become more human, this means that they will become more mobile and intelligent. They will take on more human attributes such as the ability to react to spoken words or written instructions and to reply in a way that seems more natural. The results will be a user interface that will appear more human, even if the underlying computer program is not artificially intelligent.

The outgrowth of a more human interface is that the interface assumes a personality. This could mean an interface that gets in the way more than it helps like MicroSoft's BOB. But, if it is done correctly, this could make computers and Web sites easier and more natural to use.

BOB

Microsoft's BOB was an attempt at social interface as a front end to Windows. BOB was a short-lived market failure, but the concept lives on in the Intelligent Assistant in Microsoft Office. Office uses an interface with characters like an animated paper clip or friendly dog. You can ask the character questions, or you can click it out of the interface. These techniques will be used in more computer systems over the next few years. The characters sometimes called bots or agents will probably become common in many Web sites.

Natural-language PCs

Natural-language processing is the ability to respond to a question that is phrased the way you would phrase it to another person. One Web site that uses a natural-language interface is Ask Jeeves, which has a knowledge base of 7 million answers. Unlike most search engines, you ask Jeeves a question and it gives you a list of specific questions it knows the answer to. One of these should help answer your original question.

Chatterbots

A chatterbot is a virtual character that you can ask questions of and get answers for a particular knowledge area. For example, a simulated psychiatrist could play back words and phrases based on a rudimentary analysis of the input. This demonstrates a simplistic form of artificial intelligence and some chatterbots can seem human at times and useless at others.

Chatterbots such as used by Neuromedia, in its Neuro Server Engine, provide a demonstration of a virtual customer service representative. This chatterbot exists as a box into which you type text. Customers can talk to it and get answers to their questions most of the time, cutting down on the time they spend on the phone talking to a human representative.

The chatterbot does not reason, it only matches patterns. Another chatterbot is used by the Big Science Company. The chatterbot called Andrette replies to your questions and shows different pictures of the character.

When natural language or a chatterbot is to control an application, the result is sometimes called a social interface. NetSage uses several of these, including its social Intelligence Server. NetSage uses an Animated Intelligent Agent and NetSage's languages allow companies to develop systems that act as sales or support representatives.

The Social Intelligence Server has been used to develop the Office Assistant in Office 97 and the voice user interface in General Magic's Portico virtual assistant. This telephone assistant allows you to remotely send copies and play phone messages. Another voice actuated business service is Webley's Systems Webley Service.

Voice Recognition Software

Another example of a more human interface involves voice recognition, which has been evolving at an exceptional rate. Products like Dragon NaturallySpeaking and IBM's ViaVoice are popular in continuous-speech recognition applications. Many limitations of these products will soon be overcome. One area of achievement is towards eliminating the time users must spend training the software to recognize their voices. These programs are becoming speaker independent and IBM's speech development department has demonstrated software that you call over the phone. It answers and directs your questions based on your response, without any training required.

Telephone-based systems have also been targeted by the MIT Laboratory for Computer Science. Their speech understanding systems are designed for specialized functions such as weather reports. Internet access lags telephone use, so voice is a way of bringing out applications to more people. The Jupiter system uses speech understanding to answer questions about the weather over the phone.

Emotion is another area that computers are trying to understand. MIT's Media Labs' Affective Computing Group has developed a system where physiological sensors are attached to the body and cameras record facial expressions so the computer can monitor reactions. The program will then adjust to react to the sensor and camera inputs. For example, if the user seems confused by a set of instructions, the program would repeat them or offer additional explanations.

The computer itself may also express emotion. MIT's Bruzard is an interactive animated 3-D character that uses facial expressions to react to questions. In the future, Bruzard could be combined with a chatterbot to create a more human interface.

Microsoft Research is combining these concepts in a project called Flow for virtual meetings. The scheme is to represent a traditional meeting as much as possible, even though everyone might be in different locations.

The combination of animation, natural-language processing, voice recognition and voice synthesis are moving towards user interfaces that will seem more natural.

Networks in the Future

In the future, networking will be more invisible and universal. There will be faster-wired networks at home and in the office and more high-speed services connecting everything. The areas of networking technology in growth modes include faster access and LANs will continue to grow and become a part of the utility system.

The benefits of being universally connected include more efficient utilization of resources. Industry is now deploying the world's largest virtual network as an increasing number of contacts for both buys and sales take place over the Internet.

The modern automobile is becoming a network of processors and sensors (see Table 5-1). Tollbooths scan our cards and automated fuel pumps can bill us quickly and efficiently. The connected world tries to become even more adaptive and helpful.

In manufacturing, this connectivity allows just-in-time delivery of raw materials, shipping of finished products, access to funding and management. Improvements in connectivity will reduce the need for

Table 5-1. Automotive Control Trends

- Front touch-screen LCD for car controls, e-mail, navigation system, and phone
- Rear flat-panel display for Web surfing, games and movies
- Seeing-eye cameras and embedded roadway transponders for steering
- Radar-based adaptive cruise control for maintaining a safe following distance
- Smart sensors in run-flat tires eliminate spares

energy as products and services are delivered on-line.

Networking progress continues in wide area networks, the Internet and corporate Intranets. New developments include Internet2 and other projects to link industry, government and universities.

One part of Internet2 is known as Abilene. It uses over 10,000 miles of fiber-optic cable and operates at a 2.4-Gbps with a growth potential of 10-Gbps. More than 70 universities and research facilities are connected to Abilene.

The Internet Protocol (IP) version 4 was heading for a crash because of address saturation, but the Network Address Translation (NAT) saved it by relieving the pressure on the number of available IP addresses. IP does not tell you enough about the data it transports.

The Internet needs to have greater numbering capacities for more nodes and more capacity to set priorities for different categories of data. Version 6 should provide these features for the increased loads of Internet and Intranet traffic.

There has been a massive increase of digital subscriber line (DSL) connections for businesses. Cable modems will be used for additional home subscribers. Both techniques will run into limitations on distance and the cost of installing new copper lines.

Early in the millennium, there will be fixed-point wireless with microwaves to carry high-speed data. It will be employed in many rural areas. Cable modem, DSL and wireless will all find their application areas.

The portable wireless area will continue to explode. Cell phone-based digital technologies will compete with those meant only for data. Competing technologies for browser-equipped phone connections will clash while creating more features and lower prices. In the near future Internet-connected telephone/pager/browser devices will become widely used.

Web Advances

The Web was originally designed to display text. Later, graphics were added and the Web grew quickly. In the future, Web sites will be more intelligent about their users, their content and the content of other sites. New Web approaches include personal interfaces, bots and search techniques.

The Web's most noticeable trend is toward personalization. We have seen personal portals such as My Excite, where users customize the site telling which categories and information they want.

More Web sites are becoming aware of their visitors and customizing the information they provide. These sites track visitors by giving them log-on names and passwords or cookies that are stored in the browser. Web site cookies are files that store information on visitors to that site.

The next step is for a Web site to customize its information without you expressly telling it what you want. Future sites will track what you tend to look at and offer it to you without your intervention.

Collaborative filtering is another technique that makes sites appear smart. Software from Firefly Network (which was acquired by Microsoft), Net Perceptions, and Andromedia match users and sites for products and services. These personalization services compare your web uses with others.

DirectHit uses a variation of this for Web searching. When you are doing a Web keyword search, Direct Hit points you at the site that the most users have visited. This type of popularity technique is used by sites like HotBot and ZDNet.

Another trend is the use of bots or agents. Bots, which is short for robots are tied to Web sites. They query other sites to help you find information which can include prices on products and services. In the future, you may have your own agents that reside in your computer.

IBM's Almaden Research Center has been working on software for providing Web Browsing Intelligence (WBI). These programs are similar to agents, but they act as intermediaries in the background to customize and enhance the user environment. They might automatically retrieve and install software you need to view a Web page.

HTML and XML

XML (eXtensible Markup Language) is emerging as the technology to allow web sites to understand each others content. HTML provides the language to define the layout of text and graphics on a web page. XML is not a presentation language, like HTML, but a way of describing data.

XML will enable more precise searching and let businesses share data more efficiently. The Web, which was built on HTML, will be rebuilt on XML.

Like HTML, XML grew out of the Standard Generalized Markup Language (SGML). SGML is a metalanguage, which means it is a language that lets you create other languages. SGML allows various groups in industry to create tags, elements, and attributes called Document Type Definitions (DTDs) specific to their applications.

HTML is a specific set of tags designed to display Web pages, but XML is a simpler metalanguage designed to allow groups to easily come up with standard tags to describe the contents of specific types of files.

XML is being utilized as the base for standardized tags for financial transaction, vector graphics and multimedia applications. The Open Financial exchange (OFX) specification provides a common set of tags to allow users to manage finances and pay bills on-line.

XML is a standard, so most software can be adapted to work with it. The future of XML is to allow applications of all types to exchange data on the Web intelligently. XML is likely to be used in other ways as well. In Office 2000, Microsoft is supporting XML as a file format, so Microsoft Word or Excel documents and files could be viewed in any XML-complaint browser.

Web Searching

Another area XML may improve is Web searching. Since XML tags describe information in a document, the descriptions themselves can be searched. This reduces searching time by letting you be more specific.

The Aeneid search engine uses XML tags and sophisticated text retrieval to find and correlate related information at multiple Web sites. Aeneid personalizes Web searches and allows the information to be analyzed using XML.

Most of today's search engines look for text. As the Web grows and more types of different media is available, the ability to search through this array of information becomes more difficult. Virage's VideoLogger is a product designed for coding and accessing video content. Virage's software allow users to search through video files for specific spoken words and phrases.

IBM has been developing a technique called Query By Image Content (QBIC). This should allow users to organize and search a database of images using colors, shapes and textures.

Web user interfaces will also be changing. Inxight's Hyperbolic Tree works as a set of hypertext links that allows a user to navigate spatially with a mouse. The Hyperbolic Tree references the previously read Web links as if they were on a globe. One might be located near the equator while another link might be closer to the north or south pole.

Natrificial has also been working on a product that organizes Web-based information with a spatial interface. The Brain is similar to Hyperbolic Trees and is designed to act as a medium for sharing content over the Internet without the need for a dedicated Web site to store the information. The Brain creates associative structures to relate a group of files to one another, just as a human brain associates pieces of information.

These hyperlinks are used to create connections between graphical tokens called thoughts. Each thought can hold a file or link to any other thought or level.

Intelligent Devices

Many types of devices with build-in intelligence use embedded processors like StrongARM and DragonBall and operating systems such as pSOS and VxWorkx. The pSOS operating system is being used in personal digital assistants (PDAs), cell phones and Web PDAs. Microsoft has demonstrated the Hermes Internet telephone, which runs Windows CE.

There are 120 of these chips from about 40 companies which are positioned to reinvent computing in the new millennium. They are used in ovens, digital phones, automobile control systems and other products

(Table 5-2). The Palm family of hand-held devices uses Motorola DragonBall EZ chips.

Tables 5-2. Device Trends

- Samsung CDMA watch/phone is a small wireless phone
- Internet refrigerator from Frigidaire and ICL has Internet access and bar-code scanner to track contents
- Ericsson R380 mobile phone runs on EPOC OS, with voice dialing and Internet access

Some low-cost embedded processors compare with Intel Pentiums, except that the embedded processors cost less and consume less power. Stanford University's School of Engineering believes that embedded chips may be first to offer multiple processors on a single chip.

IBM has a prototype of its cyberphone. This is a combination cellular phone and computer with wireless Web access. It has a small built-in display which uses a mirror to enlarge images to the same size as they would appear on a desktop monitor.

IBM is also working on a super-high-resolution wrist watch that can provide easy visual access to data. Wrist phones are available from Samsung with voice-activated dialing.

Symbian is a consortium of Ericsson, Motorola, Nokia, and Psion that developed the EPOC operating system for Internet devices. Symbian is developing secure wireless devices and Web surfing phones which are souped-up cell phones that provide access to the Internet.

The Ericsson R380 GSM mobile phone will run on EPOC and include a built-in modem, PDA, touch screen, calendar, address book, e-mail, Internet connection, handwriting recognition, voice dialing and voice answering.

Smarter microprocessors are also being used in refrigerators. A prototype refrigerator from Frigidaire Home Products and ICL, uses an Intel microprocessor along with 32MB of RAM, an Ethernet connection, flat-panel touch screen, bar-code scanner and Microsoft Windows. This refrigerator is connected to a phone line for dial-up Internet access or to a home network that is connected to the Web.

Communication Options

Sun Microsystems and some manufacturers of cellular phones, printers, and hand-held computers have developed Jini, which is a Java-based technology for device communication. It plants Java code on digital devices so they can be organized into groups. Devices with Jini can join a network and automatically configure themselves.

Jini competes as a common language for digital devices with Microsoft's Universal Plug-and-Play. When you connect a device to the network, the device automatically acquires a TCP/IP address and using a discover protocol based on HTTP, it announces its availability to other devices on the network.

Universal Plug-and-Play is one part of Microsoft's Millennium OS project, which aims at a network of devices that can allocate resources efficiently to accomplish each task. Universal Plug-and-Play provides a way for devices to communicate while Millennium takes these devices and makes them look like one computer. Millennium is slated for home use while Windows 2000 is designed for business use.

Motorola is a leading embedded-chip manufacturer for smart appliances. Motorola and MIT's Media Lab have started the Motorola DigitalDNA Laboratory, where future smart appliances are developed. DigitaDNA refers to the concept of linking various smart products. One of these is a smart treadmill that checks your pulse and automatically adjusts the treadmill. Other concepts in work include clothing with labels that informat the washer which cycle to use and doors that open for specified people or pets.

Smart Software

Most software is not very smart. The majority of applications used today only react to commands given them. As computers become more powerful, software capabilities also grow and future software may be much smarter.

Software that appears to understand what it is told has been around since the 1970's. In 1979 an early game called Dungeon was developed at

MIT. It allowed users to type in phrases such as "Open the door." Dungeon would react and describe what happened after the door was opened.

Other programs such as Lotus Notes can interpret a wide range of natural-language commands. Lotus Notes allows you to type phrases like "Schedule a meeting for next Monday at 9:00 with Engineering development." The program will make the proper entry in your calendar with the correct contacts. These types of applications appear to understand what you tell them as long as it is related to mail, calendar and contacts.

Natural-language interfaces are a sign that software is getting smarter and becoming more human. The speed capabilities of computers was illustrated in Deep Blue's 1997 victory over chess world champion Garry Kasparov. Deep Blue is a powerful 32-node IBM RS/6000 SP computer and was able to calculate hundreds of millions of positions per second to decide on the optimal move. Kasparov only could examine a few positions per second.

The processing power that lets Deep Blue simulate millions of possible chess outcomes is being used to perform time-consuming tasks such as the simulation of weather patterns and other complex physical models.

Computer software is also beginning to use human-like pattern recognition which is typically referred to as neural net techniques. Neural nets learn from experiences. In the future computers will aid doctors in analyzing a patient's condition and assist them in creating new drugs right in the office.

Future of PC Software

Future personal computer software will probably replace the virtual desktop in Windows with the Webtop. You may soon use Windows occasionally, but living in a Web browser as the Webtop takes over from the desktop.

In the near future you will plug into the Webtop wherever you go. It will know who you are, what applications you use, and how you want your interface to look. No more configuring of your personal computer for communications will be needed. Communications will be easier, as everyone logs into one big network. The software industry needs to change fundamentally in the same way the microchip changed the design of circuits.

Bayesian Networks

Some intelligent concepts are being introduced into PC software. In Office 97, Microsoft introduced the Office Assistant. This animated icon watched what you were doing in your word processor or spreadsheet program and offered tips and answered questions to help you with the program.

The concept behind tools such as the Assistant comes from Microsoft Research's Decision Theory and Adaptive Systems Group. This group has been involved in Bayesian Networks. These networks are interconnected on the probability of concurrence among different actions and conditions. For example, in Microsoft Word the program follows the actions by the user, infers what the user is trying to accomplish and makes suggestions through the Assistant.

These initial examples are relatively simple, but Bayesian Networks can be much more sophisticated. Bayesian networks have been used as a way to improve processes. They are also finding their way into Portal software for hypertext systems. This type of modeling of the decision-making process will allow software in the future to help you with complex problems. There will be a gradual increase in software intelligence that help you do more.

Personal Slaves

At Microsoft they see a future where the PC not only transcribes speech but understands what you say. Microsoft predicts a future without application programs. You would tell your PC what you want and let it sort out the details.

The main issue about today's software is that it requires too much tinkering. In 5 or 10 years software will know when its sick and what it needs to get well. Future versions should offer troubleshooting wizards that help find hardware conflicts and printing and modem problems.

When you have a problem, a wizard will ask relevant questions to help you figure out what's happening. In the future, software will look for conflicts, make sure drivers are up-to-date and when a fix is necessary, ask if you want to go on-line and provide a patch. Later on, it will search for the changes it needs with no intervention from you. Future software may even watch what you are doing and step in when you are having trouble. This software will work when you need it as a personal slave in the computer and allow easy collaboration over the Internet.

The PC now handles many tasks that once were the sole province of large computers. The PC of the future will have terabytes of storage. Displays may be flexible and you may unfold them from your pocket-like cell phone. Voice recognition may not completely replace mice because of privacy concerns.

You will need lots of connections and they will be in the walls and the floor. What may go away is those wires to the phones, the keyboard and the mouse. These may all go wireless.

PCs should be more reliable and customizable. They will need to free us from maintenance hassles, compatibility glitches and an ever-spiraling rate of obsolescence.

The technology should provide advances in processors, storage, displays, and all the supporting technologies needed to make the PC more useful.

Processing Power

The innovation for the years ahead in silicon fabrication and CPU architectures will produce a billion transistors on a chip by 2011 or sooner and computing devices will be much more powerful. Intel's Pentiums and AMD's faster K6 chips are tuned to handle graphics better. Intel founder, Gordon Moore, predicted that processing power would double every 24 months. Intel's Pentium II has 7.5 million transistors. As the trend continues, Intel processors should contain 50 million to 100 million transistors by the first decade of the 21st century (see Table 5-3).

In 1965, Intel Corporation cofounder Gordon Moore predicted that the density of transistors in an integrated circuit would double every year. His observation was called Moore's Law and was later changed to every 18 months. Moore's Law has been relatively accurate for over 30 years. Transistor density as well as microprocessor performance tends to follow Moore's Law. The cost of memory also continues to plummet (see Table 5-4).

In 1996 Intel predicted that by 2011, they would have a microprocessor with 1 billion transistors which runs at 10-GHz. This would be done using 0.07-micron semiconductor processing technology. The chip

Table 5-3. Intel Processor Growth

1999	Pentium II - 450-MHz Pentium III/667 (Coppermine) - 0.25-micron to 0.18-micron process, bus goes from 100 to 133-MHz
2000	Pentium III/700, Celeron 500+
2000-01	Willamette—1-Gigahertz
Mid-2000	Merced—First IA-64-chip
2001	McKinley—Second-generation IA-64 double the speed of Merced (over 1-GHz)
2002	Madison—Third-generation IA-64 Deerfield—Price/performance IA-64
2002-03	Change to 0.13-micron process with copper interconnects
2011	10-GHz, 1 billion transistors 100 billion operations per second

Table 5-4. Microchip Trends

Year	*Megabytes Per Dollar*	*Transistors (thousands)*
1988	0.1	1,000
1992	1	2,000
1995	10	10,000
2000	100	50,000
2002	1,000	100,000

would be able to handle 100 billion operations per second.

In the future, doubling the transistor count may require more than 18 months, due to major changes in chip design or fabrication technology. Chip logic is becoming more complex and requires longer design and validation times.

Fabrication Technology

Fabrication technology must improve with each process generation, such as the change from 0.25-microns to 0.18. A critical part of the process is photolithography, where short-wavelength light sources are focused with precision lenses through the transparent masks containing the circuit details. This exposes the photoresist on a wafer's surface, which is chemically removed leaving the details of the circuit pattern on the wafer.

Light sources and optics must evolve together. Intel's 0.18-micron Pentium III chips use the same 248-mm wavelength deep-UV light source as 0.25-micron Pentium II and Pentium III chips. The move to a 0.13-micron process should be coupled with 193-nm wavelength light from excimer laser sources.

Beyond 0.13-microns would be a 0.09-micron process, which would use 157-nm wavelength excimer lasers. The next step below 0.09 is major in terms of technology and manufacturing processes. This would be the 0.07-micron process for the 2011 processor.

The level of photolithography will probably require extreme-UV (EUV) light sources. EUV has a wavelength of 13nm, which has the potential for etching much smaller transistors. The problem is that there are no known transparent mask materials that will allow these short wavelengths to pass through. New reflective lithography processes and optics would need to be implemented.

As the number of transistors increases on a chip, transistor interconnect wires become smaller and closer together. This increases conductor resistance and capacitance which adds to any signal delays.

To reduce resistance and shrink the interconnect line widths, copper can be used to replace aluminum as the interconnecting metal. Copper is being used in IBM's Power G3 chips and AMD's newer ships. In the future Intel CPUs in the 0.13-micron class should use copper interconnects.

Physical Limitations

Power and heat management become major problems as transistors continue to shrink. The gate oxides become very thin in order to maintain the required transistor switching speeds. This forces low voltage operation in order to maintain their structural integrity. Intel microprocessor chips 10 years from now may operate at less than 1 volt and could consume 40 to 50 watts of power. This implies 40- to 50-amp currents. Evenly distributing this amount of current within the chip and dissipating the heat generated are both potential problems.

Current silicon fabrication methods may hit physical limits by the year 2017, which many have predicted. At this point, it will not be possible to build usable transistors any smaller.

By that time, research into areas such as molecular nanotechnology, optical or photonic computing, quantum computing, DNA computing, chaotic computing and other areas of research may be more common.

Molecular nanotechnology involves the manufacturing of a electronic chip with tiny programmable assemblers to move individual atoms as discussed in Chapter 3. This technology would move devices into the nanometer range rather than tenths of microns as it is now. An entire computer made this way would be about the size of a sugar cube.

Photonic computing involves using beams of light for computing functions. Quantum computing technology involves the study of electrons floating in liquid helium as computing elements.

DNA computing is patterned after the way DNA stores information very efficiently. The DNA strands can find each other among thousands of other strands. Synthetic molecules called lipsomes are used for searches that requires massive parallel processing. It is also called biomolecular computing. Chaotic computing attempts to apply chaos-science theory to the management of computer networks.

There are apt to be major changes in the design and manufacture of microprocessors and the way they perform computer operations. This means that as fabrication technologies sustain enormous changes in the coming years, the microprocessor architectures that are used today including logic designs, instruction sets, registers and external interfaces are bound to change. Expect to see more complex 32-bit x 86 processors from AMD, Cyrix, Intel, and others in the near future.

Intel and Hewlett Packard have developed the EPIC (Explicitly Parallel Instruction Computing) instruction set technology. This is departure from the Intel x86 architecture. The 64-bit IA-64 architecture is the first instruction set to use EPIC.

The forthcoming Merced processor will be the first actual IA-64 implementation. Intel believes 64-bit processing will be the mainstream technology in about 10 years.

An important objective for AMD is to get as much fast memory as close to the processor as possible and to reduce latencies to I/O devices. Future CPU chips will have faster and more direct interactions with main memory, graphics and lower bandwidth streaming devices. There will also be a trend toward PC-on-a-chip devices.

Chip Multiprocessors

Chip Multiprocessors (CMPs) have multiple processor cores on a single chip. These devices are expected to proliferate over the next decade.

There will be more multithreaded applications and multitasking that take advantage of these architectures. These multiprocessing designs may delay the shift to more exotic computer designs.

The embedded-CPU market would be the first area for CMPs. CMPs would then move to workstations and servers, although memory bandwidth for the multiple cores could slow this trend.

Future Buses

Even the fastest processor can seem slow if your PC's other subsystems and data pathways are not up to high speeds. Intel and other chip providers are assaulting the throughput problem by improving the buses the data must move through.

Intel's 440LX AGPset chip set introduced the Accelerated Graphics Port, which is a faster than PCI bus between your CPU and graphics coprocessor. AGP also links the graphics adapter to system memory for storing the large texture maps that are used in 3D applications.

New system board chips will also support faster system bus speeds of 100-MHz or greater. Increasing the bus speed lets the PC take advantage of faster memory types. The 100-MHz synchronous DRAMs as well as Direct RDRAMs provides better performance and lower costs.

Faster Connections

In the future you will be able to connect more devices to your PC, which will exchange data with them at much higher speeds. This is because parallel and serial ports will be replaced by high speed buses like the Universal Serial Bus (USB) and the IEEE 1394 high-speed serial interface (FireWire). USB provides transfers at up to 12 megabits per second which is about 100 times faster than most standard serial ports. FireWire starts at 100 to 400-mbps and should rise to 1-gigabyte per second in about 10 years.

USB is used to connect low-bandwidth devices such as mice and scanners in many new PCs. FireWire will link external DVD and CD-ROM players, digital video, still cameras and hard drives.

Wearable Electronic Displays

A plant supervisor may soon be able to receive vital information and see multiple areas of the plant through special safety glasses. A technician may be able to probe the circuit and see the output of an oscilloscope in the same view, eliminating the need to look back and forth. Wearable electronic displays can be mounted on or integrated into most frames or safety glasses, letting users obtain and view data while wearing normal vision aids. These new eyeglass displays use a biaxial scanning mirror to raster-scan modulated laser light onto a screen. When the light modulation and mirror motion are synchronized, an image forms. Single crystal silicon is used for the mirror.

In one version the housing of the conversion electronics is separated from the display by a 4-foot cable. The display has a 320- by 340-pixel resolution and a 60-Hz refresh rate. The unit has a head-supported weight of less than 50 grams and a field of view of 12 degrees diagonal.

Another version integrates the see-through or see-around display optics directly into the lens and includes a proprietary lens system that relays the image to the eye and provides prescriptive correction. This version has a 10 diagonal field of view and a head-supported weight of 110 grams.

The demand for wearable electronics like these is expanding. Researchers at Boeing Company have designed a pair of computerized eyeglasses that allow the wearer to interact by voice or video with a host system. The device could enhance maintenance and service procedures using features such as eye movement detection, speech recognition and visual input capability.

The eyeglasses are based on traditional safety glasses and provide the wearer with audio and video interaction to a host system. The glasses could contain sensors to detect eye movement, a microphone to handle speech input for speech recognition and a miniature camera to allow visual inputs. The ear pieces could contain the power supply, neural net and other processors. Audio feedback could also be part of the combination.

The user would be able to communicate with the device via natural speech, eye movement, or even mental feedback with the appropriate detectors for alpha waves or other brain signals. Feedback would be in the form of a retinal display projected into the eye or speech synthesis amplified from a small speaker.

A high-resolution digital video camera would be located on the nose bridge. The camera would have to include some type of electronic focus and iris, as well as enough pixels to allow software to scale the image up (magnify) without losing details in blocks of pixels.

The camera would provide a high-resolution view to central maintenance to communicate what is being seen to support staff or intelligent agents (bots). The camera could also provide visual assistance by sending magnified, infrared, or thermal images directly to the retinal display system.

The eyeglass stems could hold the power supply and the local processor. A rechargeable battery takes up about one-half this area. To handle the image and audio processing, the computer unit would contain a Neural-Net Processing Unit (NNPU) as well as conventional nonvolatile storage and CPU arrays.

Located at the ends of the ear pieces, the communications system would provide all communications including voice, wireless network, and mental feedback. The circuits also could monitor the vital signs of the wearer in the event of an emergency.

Images from the camera or communicated from the host over a wireless link would be fed to the retinal display system. The system projects from the side of the safety glasses into the lens, reflecting back to the user's eye.

This type of system should be able to provide a resolution of about 3000 pixels by 5000 pixels in each eye. The system also monitors eye movement for certain types of user input. For safety reasons, outer peripheral vision would not be projected during use in an operational environment.

MHA Technology

Magnetic haptic array (MHA) technology can provide tactile feedback in almost any angle and position. MHA technology uses gravitic forces rather than magnetism to produce force vectors on the user's hands and forearms. The magnetic in MHA refers to the use of weak magnetic fields for position sensing.

A pair of magnetically sensitive gloves allows the MHA to project gravitic forces. The user would be able to perform mechanical procedures with full tactile feedback on the hands and forearms.

Displays

New technologies are also being developed to replace standard displays such as cathode-ray tubes (CRTs) and liquid crystal displays (LCDs) which may be bulky, or hard to see and consume too much power.

IBM has developed a new flat-panel technology that provides four times the resolution of Super XGA (1,280 x 1,024) which is the best LCD display currently available. Instead of using the display materials of molybdenum and tungsten, these new displays use aluminum and copper, which are better conductors and provide higher resolution.

The Roentgen 16.3-inch display has a resolution of 2,560 by 2,048 pixels which is over 5 million pixels with 200 dots per inch. It will initially be available for medical imaging, but should soon find its way to laptops.

Xerox PARC has been working on the ultra-thin Gyricon display, which uses millions of plastic balls enclosed in a pocket of oil within a sheet of rubber. Each ball has a white half and a black half and carries an electric charge. The black side creates black dots, and the white side creates white dots, much the way photographs are displayed in a newspaper.

The balls can be locked to provide long-term storage of an image. When the charge is changed, a new image appears. The resolution is about 400 to 600 dpi, which is close to laser printer quality. Most desktop displays are about 75 dpi. Xerox hopes the technology will be flexible enough to create electronic books, newspapers and even wallpaper.

Xerox is also working on DataGlyphs, which is a technology for printing machine-readable data on paper documents. The glyphs are embedded in the documents. DataGlyphs can encode text, data, or graphics on plain paper. This can then be scanned into a computer and used for bringing up a Web site or printing documents. This process is also known as smart paper.

E Ink is an electronic ink technology that uses electrophoretic processes developed by the MIT Media Lab. E Ink is an ink-like material that can change color using an electric field. E Ink may appear in changeable traffic warning signs and store displays.

Both E Ink and Gyricon may be used for creating electronic books. Most of these machines, which can download material from the Web, use conventional LCDs. The E Ink or Gyricon processes provides an electronic book that is thinner, lighter and more readable compared to a LCD.

A related electronic book concept is underway in a project called Folio at the Interaction Design Studio at Carnegie Mellon University's Institute for Complex Engineered Systems. Folio's display is made up of 8 hinged, full-color polymer LCD panels that fold up to fit in a pocket. Unfold it once and it becomes a writing pad, unfold twice and it becomes an electronic book or Web browser. Unfolding it completely and it is large enough to display maps or work as a large-scale PC screen.

As digital characters begin to look and act more human, actors in movies are becoming increasingly digital. Automated response bots can simulate human conversation or take human form in computer programs. The Internet will become an increasingly larger source of video content and it will produce high-quality digital video.

Digital Characters

Animated characters will continue to look and act more real. Much of this visual realism is due to 3-D hardware capabilities of the computer operating platform. The Intel Pentium chip has an instruction set called Streaming SIMD Extensions (SSE). It is designed to speed 3-D geometry calculations in applications that use the extensions. Many SSE instructions match those found in AMD's 3DNow!

Many future graphics chips will perform geometry calculations on-chip. This will produce more finely detailed graphic images and characters.

These new technologies should allow a broader, richer set of interactive behaviors that work well in low bandwidth environments like the Internet. Some of this technology has been used in robotics and real-time process control. These systems build on specific skills and behaviors, allowing a set of reactions to user input that are more realistic in appearance and also more intelligent.

ATI has several graphics cards that can produce MPEG-2 video in real time on a Pentium III-based PC. MPEG-2 provides video that looks as good as broadcast video because it is broadcast video, serving as the video compression technology for satellite-based video delivery systems.

Stereoscopic headsets can monitor head movements and input them into the computer. These are being integrated into DirectX, which is Microsoft's standard for interfacing applications.

Digital Identity

A digital identity is that constantly growing mixture of personal information which is stored in the databases of state and municipal offices, hospitals, medical centers, insurance companies, stores, banks and federal agencies. The development of digital identities has been accelerated by the Internet. Bits of information about buying habits, reading preferences, opinions, browser bookmarks and financial data may all be available on the Internet.

As we begin a new millennium, privacy is being lost in the information given to bankers, insurance companies and other service providers. Privacy has been bartered away, for services and on-line conveniences. Privacy in the digital age means the ability to control our information. There will be a widespread trend for individual access to information control. This trend will continue toward more individual control. In the early years of computing there were relatively few digital records that were related to personal records. Today, many types of personal data are being gathered for possible use in biometric security systems.

These systems are meant to verify your identity or to allow access to buildings or facilities. These include digital brain scans, electrocardiograms and MRIs (see Table 5-5).

Table 5-5. Biometrics Indemnity Characteristics

- Individual attributes of face create a digitized mathematical model.
- Iris and retinal patterns can be scanned like a human bar-code.
- Spoken words can be matched with voiceprint for remote access.
- Geometry of hand, veins in wrist, map of skin pores can be matched.

There is a growing trend by Web-based businesses that they must post detailed privacy policies on their sites. Many of those policies are verified by TrustE, which is an independent, nonprofit consortium to promote principals of disclosure and informed consent. TrustE licensees must agree to disclose information and privacy practices that include the type of information being gathered, how the information is used and how it is shared.

Technological solutions to this problem include encryption, which has been a major vehicle to help ensure digital privacy and security issues. A variety of encryption schemes, including public key systems, have been used in many computing and Internet applications to protect digital data.

Internet Security Systems has its Internet Scanner which has an Explorer like tree-interface, but no firewall scanning. Network Associates' Cyber Cop Scanner does have firewall scanning. Other products include Axent Technologies' Net Recon and Web Trends' Security Analyzer, which was designed for Web servers, but is evolving into a general-purpose scanner.

Encryption will become more effective, transparent, and user-friendly in e-mail, browser and e-commerce applications. Most of this effort will be based in server-based encryption rather than approaches that require end-user intervention. But, end users will still be able to apply technical solutions on their own. The newer versions of Microsoft Internet Explorer and Netscape Navigator have many features for setting user profiles, managing cookies and filtering content. Secure digital signatures and certificates are handled along with encryption for limiting the amount of information revealed.

There will be further development of security architectures and approaches by groups such as the World Wide Web Consortium's Platform for Privacy Preferences.

Security levels are used to automate a user's interaction with Web sites through compliant browsers, plug-ins, or servers. Intel has developed the Common Data Security Architecture (CDSA). This is a set of application programming interfaces (APIs) designed to make computer platforms more secure.

The growth of smart cards indicates their acknowledgment as security devices. A smart card can be used with passwords or biometric data to protect access to data. The digital identity is embedded in the chip on the card. In the future, those chips will be embedded in clothes and perhaps eventually in our bodies.

Public key encryption-based digital certificates now have limited use in browsers. These will become more common in the future. They can be used by almost any program and will be a feature of most smart card identity systems.

Authentication card software includes Security Dynamics Secure ID program and freeware programs from Trip Wire, Crack, COPS and TCP Wrapper. IBM's Keyworks is designed for public-key infrastructure services including cryptography. Lotus Notes uses Keyworks. IBM also has its Network firewall for Windows NT.

Random Number Generation

Random number generators provide results that may appear to be random, but they are actually pseudo-random numbers generated by a calculation such as the Lewis-Goodman-Miller generator. This means that over a period of time a pattern will emerge in the numbers that are computer picks. That can cause trouble when these numbers are used in security schemes.

Random numbers are fundamental to computer security, since they are used to generate security keys that prevent breaking into on-line systems. Given enough time the intruder can trace a pattern and replicate a key.

Intel has developed a random number generator that uses thermal noise. This is one of the few sources of random and indeterministic phenomena. This true random number generator is then used as an input to produce secure keys.

Thermal noise is produced when an electrical resistor heats up as a current is passed through it. This is due to molecular action that gives off energy as heat.

Electronic hardware-based random number generators tend to be bulky and expensive. Intel is producing its hardware-based random number generator as a standard integrated circuit component. Intel will include the random number generator in its 810 chip set, which will be used in standard desktop PCs.

Industrial Computers

PC-based processor boards now include systems that allow PowerPC speed and performance for a wide range of industrial applications. One processor board uses the MPC8240 PowerPC chip, which is a highly integrated microprocessor built around a PowerPC core. The unit has 200- or 250-MHz processors, up to 128 MB of synchronous DRAM, up to 8 MB of flash memory, and either 512 KB of nonvolatile RAM or 144 MB of disk-on-chip memory. On-board serial I/O, Fast Ethernet, two RS-232 serial ports, and a PCI expansion connector are also included.

An Intel 82559 Fast Ethernet controller, with an integrated 10/100-Mbps physical layer device provides a compact LAN controller. The PowerPC board uses an operating system-independent boot loader for compatibility with any operating system.

The 64-bit PowerPC 750 microprocessor is also available in a single-board computer. It can run on a choice of operating systems, including VxWorks/Tornado and ISI pSOS. Along with the 400-MHz processor, the board has 128 MB of high-speed synchronous DRAM, which can be expanded to 256 MB through memory modules. A Fast Ethernet 10/100Base-TX interface is included with an RS-232 serial I/O port that can be accessed from the front panel. A 32-bit PCI local bus allows high-speed interconnections. There is a real-time clock, and three 16-bit timers. A PCI-to-PCI bridge is used to connect the computer to a CompactPCI system bus.

Other Single-board Computers

Another single-board computer uses an AMD K6-3 500-MHz processor, high-performance Ultra2 SCSI, 10/100Base-T Ethernet, and 4 MB of AGP video. The board supports voltage and software watchdogs, a processor thermometer and an internal fan-speed monitor. Standard interfaces include two USB ports, two serial ports, keyboard and mouse ports.

Rugged Computers

Rugged single-board computers incorporate the 400-MHz Motorola MPC7400 PowerPC processor. An extensive memory bandwidth is possible with an 83.3-MHz memory bus that yields sustainable memory transfer rates of 190 MBps and 296 MBps. Also included is cache support, flash capacity and 100Base-T Ethernet support for industrial applications.

All-weather workstations are available with 400-MHz CPUs. One unit has a 12.1-inch, sunlight-readable TFT display, four ISA PCI slots and a detachable keyboard and mouse. These workstations can be used in harsh environments that include rain, dust, and extreme temperature ranges.

The unit also includes a shock-mounted, 8-GB hard drive, a floppy drive and AC, DC, or battery power. A touchscreen, a solid-state drive, a PCMCIA CD drive and Ethernet are also available. The unit's shock-mounted internal parts can withstand high shock and vibration levels and it may be carried around or mounted on a panel or rack.

Other industrial computer trends include Pentium MMX-class processors with VGA-LCD interfaces, audio, 10/100Base-T Ethernet and CompactFlash card. Up to 64 MB of RAM are available. Low power consumption allows a smaller, fanless design.

A bundled Windows CE operating system, a parallel port, two serial ports, two USB connectors, keyboard connector and infrared port are included.

Pentium-based single board computers with 100-MHz bus clocks have up to 128 MB FPM/EDO (fast page mode/extended data output), 512 kB standard L2 cache memory and a socket for a Flash memory disk (DiskOnChip) of up to 72 MB. Also included are two RS-232 serial ports, a parallel port, 4 IDE interfaces, a floppy drive interface, keyboard and mouse interface.

Pentium II/III Computer Trends

The Intel 440BX chipset can be used in industrial single board computers with a 450-MHz CPU and 768-MB SRAM system memory. The cache memory is integrated in the CPU. An ATI RAGE XL AGP graphics chipset provides up to 8 MB SDRAM for graphics. I/O may include serial and parallel ports, interfaces for hard disk drives, interfaces for floppy disks, keyboard interface and DiskOnChip to 144 MB.

Cyrix-based Computers

Trends include Pentium MMX-class processors with VGA/LCD interfaces, audio, 10/100Base-T Ethernet, CompactFlash card sockets and connectors for up to 64 MB of RAM. Low power consumption leads to a smaller, fanless design and a two-chip Cyrix processor can use less real estate than an Intel architecture. Many systems are bundled with the Windows CE operating system. These single-board computers provide parallel and serial ports, USB connectors, keyboard connector and an infrared port. They can be operated in harsh environments where solid-state storage and small size are desired.

LCD Computers

Combining a 233-MHz Media GX processor, 12.1-inch flat-panel display, resistive touchscreen, 3-GB hard drive and Windows, the WebLink PC provides a platform for control applications. This computer has 32 MB of RAM, 10Base-T Ethernet port, serial port, parallel port, expansion slots, and a NEMA-4 front panel. The WebLink PC is designed for space-limited plant-floor applications.

Windows CE Computers

Industrial computers also have full-size SVGA color LCD panels with touchscreen capability running the Windows CE operating system.

These computers use a 32-bit, 190-MHz RISC processor, 16 MB of flash memory, and 32 MB of RAM. The 12.1-inch color LCD is sunlight-readable which makes the unit eight times brighter than other units. These units can be used in all-weather, outdoor, and plant-floor environments. The operator-interface touchscreen is large enough for interactive graphical use.

Industrial Software

The Allen-Bradley Panel-Builder32 has a user-friendly, configurable environment with floating/docked windows, help tools, and toolbar and workbook-mode aids that simplify system design and reduce start-up times. The software uses warning filters that allow operators to selectively shut off validation warnings. If an error is introduced during testing, a problem navigator indemnifies the problem on the validation screen. The software is compatible with RSNetworx, RSLinx, and RSLogix.

Windows-based Automation Software

Microsoft Windows NT Embedded software is designed for scalable, embedded solutions for measurement and industrial automation applications. Windows NT Embedded design and development tools can reduce the development time of automation systems. The software allows using flash or solid-state memory and supports systems that do not need a keyboard, mouse, or monitor.

Indusoft's SCADA/HMI software includes Indusoft Studio for Windows NT and CE. Application types include:

- control room/supervisory,
- operator Workstation for direct monitoring and control, and
- local interface for embedded operator interfaces.

An object-oriented database is used with math functions, report generation, archiving, alarms, interfaces for PLCs, remote I/O, and TCP/IP networking. The software can be used in embedded controls, control panels, maintenance terminals and mobile supervision.

Windows CE Software

OpenHMI lets operators and maintenance personnel analyze system performance. Use of the Windows CE operating system provides an open, PC-based computing technology rather than a proprietary system. By flattening the architecture and facilitating data flow across the enterprise, information is available on management, finance, engineering, control and logistics from one system.

HMI Development Tools

A new range of HMI tools that support Visual Basic, Visual C++, ActiveX, and Java is evolving. Programs like DataViews for Windows use a building-block structure that extends the capabilities of all tools. There is a growing need for flexible, scalable solutions that support an open architecture and open standards. The software allows interactive HMIs for the analysis, monitoring and control of process automation, telecommunications.

DataViews Express works with Visual Basic and ActiveX technology. DataViews Professional offers a C++ application programming interface (API) and a Java-class library that uses Java native interface technology. DataViews Enterprise adds a C API, a custom editor for building application-specific editors, and a custom data browser.

Visualization Software

Scalable visualization tools allow interactive Web publishing and can download applications to a variety of operating systems and platforms from a central Windows NT-based development environment. End users can benefit from the ease of use the Web-based visualization provides. The software includes run-time and Web publishing for the Windows CE, NT, and NT Embedded operating systems. Linux, VxWorks and other embedded operating systems will also be included in the future.

Control Maintenance Packages

Other software provides a systematic life-cycle approach to managing and maintaining automation control software. The software automatically compares the newly edited program with the previous copy when a program is checked back into the library. It then reports the results of this comparison via e-mail to designated users, indicating who made the changes and how the changes were made. This type of feedback helps maintenance personnel alleviate bottlenecks, avoid downtime, and simplify the maintenance of control systems. A variety of users can easily obtain information critical to the maintenance of the control environment from anywhere inside or outside the plant and with no other software except a standard Web browser.

Interbus Automation Software

Automation software combines proven software standards such as IEC 61131-3 and Windows NT into one control package. PC Worx provides programming, configuration, monitoring, and diagnostics of an Interbus system and can be used with Interbus field controllers. The package consists of System Worx for configuration, monitoring, and diagnostics. Program Worx is an IEC 61131-3-based programming tool for creating the application control program.

With the help of the Interbus system OLE for process control (OPC) server, PC Worx can exchange data with any HMI or software package that has an OPC client. Standardized interfaces such as OPC simplify the data exchange between PC Worx and other NT programs.

Ethernet I/O Drivers

New software drivers support the Linux operating system and combines TCP/IP communications, 10- or 100-Mbps performance and open protocols. These I/O systems have built-in Web servers for configuration, maintenance and operator interface. With Ethernet I/O, users can interface analog and digital I/O devices to a computer for monitoring and control applications.

Wireless Local-area Networks

The WaveLAN/IEEE Turbo system can provide wired Ethernet-equivalent data rates and improved response times for data-intensive network applications. The network is fully compliant with the IEEE 802.11 specification, which allows operation with terminals, printers, and bridges from different manufacturers. Auto rate selection automatically selects the optimal operating modes for the wireless network.

Universal Communication Modules

These modules allow ControlNet, Interbus, and Ethernet networking to a universal communication system (UCS), which provides an open-architecture, industrial, device-to-network interface for networking. The UCS client interface allows plug-and-play with all industrial networks supported by UCS master or slave modules.

Installing the appropriate network personality module into their device allows users to interface to both modern and legacy networks. The modules can be upgraded in the field and feature electrical and software independence for each supported network. UCS is capable of bidirectional communication and supports a 64-KB addressable data range.

DeviceNet Troubleshooting Tools

DeviceNet users have a tool to troubleshoot their networks. The hand-held DeviceNet Detective can quickly assess network problems. The unit has the following capabilities:

- anonymously eavesdrop on a network,
- register on the network as a slave,
- monitor voltage levels on the network, report errors and diagnostic messages,
- set node numbers,
- indicate which node numbers are being used, and
- report which nodes are masters or slaves.

Another unit called a DeviceNet Message Display uses an LCD display to provide alphanumeric messaging, operator instructions and I/O status. These DeviceNet-addressable, 4- by 20-character, back-lit units can store up to 16 messages in nonvolatile memory. The display is controlled using standard DeviceNet commands. Multiport Taps provide for the extension of a DeviceNet network by allowing more nodes to be connected to the trunk line, as compared to a single-port tap or the direct connection of a node to the trunk line.

References

Bursky, Dave, "Electronic Glasses of the Future Deliver Data, Provide Feedback," *Electronic Design*, June 23, 1997, Vol. 45 No. 13, p. 32.

Chouinard, Don, "Windows CE: the Value's in the Details," *Instrumentation and Control Systems*, Vol. 72 No. 6, June 1999, pp. 41-47.

Daflucas, Mark, "The Road to Mass Customization," *Industrial Computing*, Vol. 17 No. 6, June 1998, pp. 18-19.

Derfler, Jr., Frank J., "Networks Will Be Ubiquitous," *PC Magazine*, Vol. 18 No 12, June 22, 1999, p. 114.

Gottesman, Ben Z., "Software Will Get Smarter," *PC Magazine*, Vol. 18 No. 12, June 22, 1999, pp. 128-131.

"Industrial Computers," *Industrial Computing*, Vol. 18 No. 12, November 1999, pp. 12-13.

"Industrial Software," *Industrial Computing*, Vol. 18 No. 12, November 1999, pp. 21-23.

Kinnear, Dave, "Low-power Processors Create Performance Trade-offs for Windows CE," *Wireless Systems Design*, Vol. 3 No 6, June 1998, pp. 18-22.

Klein, Mike, "Realizing PC-based Control True Potential," *Industrial Computing*, Vol. 18 No. 9, August 1999, pp. 15-16.

Labs, Wayne, "Software: Open Source OSs, Objects, Web-based Communications Challenge Status Quo," *Instrumentation and Control Systems*, Vol. 72 No. 7, July 1999, pp. 25-49.

Levin, Carol, "Little Devices Will Think," *PC Magazine*, Vol. 18 No. 12, June 22, 1999, pp. 120-123.

Levin, Carol, "Safety in Random Numbers," *PC Magazine*, Vol. 18 No. 12, June 22, 1999, p. 30.

Miller, Michael J., "Computers Will Be More Human," *PC Magazine*, Vol. 18 No. 12, June 22, 1999, pp. 104-113.

Rupley, Sebastian, "The Web Will Be Smart," *PC Magazine*, Vol. 18 No 12, June 22, 1999, pp. 116-119.

Rupley, Sebastian, "You'll Look at Computers in a Whole New Way," *PC Magazine*, Vol. 18 No. 12, June 22, 1999, p. 134.

Stamm, Nick, "Moore's Law Will Continue to Drive Computing," *PC Magazine*, Vol. 18 No. 12, June 22, 1999, pp. 146-147.

Chapter 6
Artificial Intelligence, Fuzzy Logic and Control

When a machine performs an act or task that seems to have required intelligence, it is called an artificially intelligent act. Intelligence is defined as both the ability to learn from experience and the ability to adapt to a surrounding environment. The field of artificial intelligence (AI) involves the use of machines that try to duplicate various acts of intelligence that humans can perform.

Artificial intelligence is concerned with the development and deployment of machines that mimic human behavior. It includes the following areas:

- expert systems,
- robotics,
- vision,
- natural language processing, and
- artificial neural networks (ANN).

Of these, expert systems and ANNs will have a greater impact on HVAC control.

The study of intelligence in humans and animals helps us understand better how to build intelligence into machines. Also, studying machines that demonstrate intelligence helps us to understand humans and animals.

Intelligence can be demonstrated in several ways:

- learning from experience,
- adapting to changes in an environment,
- receiving and processing vision, speech, and other inputs to obtain specific information,

- making decisions based on the input of information, and
- generating actions from these decisions.

Learning is one of the more obvious areas of intelligence. A less obvious area involves built-in or inherited intelligence. Built-in intelligence may be stored in computer memory.

HVAC control depends on sensor input and interpretation. In the future these sensors may include speech recognition, vision, and touch.

Artificial intelligence has been important in advances involving such areas as expert systems, vision processing, image understanding, speed recognition, speech synthesis, and natural language understanding. Some of these areas are already finding applications in commercial computer products including the Internet.

Search engines like Excite, Infoseek and Lycos allow you to enter queries in plain English. You can type a question as a complete sentence in the Search Form text box: "Where can I find temperature data for northern Florida?" In Lycos, you type in the question in the search form text box and select Natural Language Query.

Much of artificial intelligence is occupied with trying to teach machines to think, but artificial intelligence is also concerned with how human beings think. The use of intelligent computer tutoring systems allows the computer to tell a user that an answer to a question is wrong and to pinpoint where the thinking process was in error.

Levels of Artificial Intelligence

Researchers have defined three different levels of intelligence: alpha, beta, and gamma. The alpha level of intelligence is defined as a nonlearning intelligence. At this level, past experiences do not enter into the decision making process. An example of this level of intelligence are the simple hand-held electronic chess-playing games. This game will always repeat its moves.

Beta-level Intelligence

The beta level of intelligence has an elementary learning capability. This level of intelligence can learn from its mistakes and past failures are used in its decision making. Beta-level intelligence is used in chess playing robots that will not choose the same losing move twice. These machines try to learn from their past mistakes and experiences. A domestic robot that is required to work around the home may need to learn its way around the house and exhibit some beta-level intelligence.

Gamma-level Intelligence

The gamma level of intelligence uses the ability to make generalizations from past experiences. This is also known as heuristics. Computer programming attempts to achieve this level of intelligence have not yet been very successful and this area has not been very well defined. Gamma-level intelligence requires that averages and other statistics be calculated and used for general conclusions. Even among humans, drawing conclusions based on statistics often results in different outcomes.

Built-in Intelligence

Many hand-held electronic games use some built-in intelligence. This intelligence is placed in their memories during the manufacturing process. They do not learn, but they may use a random number generator to choose different moves to change the outcome of the game. The use of these different moves gives the appearance of a variable style of playing the game.

Game playing machines generally use mathematical formulas to calculate the value of an alternative move, and they often use memorized opening and ending positions. This technique is also used in many industrial robots.

The human nervous system uses built-in knowledge to control functions such as heartbeat, breathing, temperature, digestion, reflexes, the ability to learn and the movement of information between the brain and the senses and muscles. Without this built-in intelligence we could not survive.

Many automated devices, from clocks to windup toys and complex electronic devices depend on built-in intelligence for their functions.

In a mechanical windup clock, the intelligence for keeping track of the time is built into the gears. The digital electronic clocks have their intelligence built into the electronic circuitry. A clock can be though a special-purpose machine whose task it is to keep the correct time. Some electronic clocks can even accept spoken commands asking what time it is and respond through a speech synthesizer with the time.

Turing's Test

Alan Turing was a British mathematician who proposed a classic test for machine intelligence. Turing's test involves a room with two computer terminals. One of the terminals is connected to a terminal in another room, with another human at the keyboard. The other terminal is connected to a computer in another room, with an artificial intelligence program loaded in its memory.

Now, use a keyboard to carry on conversations with the two terminals. If at the end of 15 minutes, you cannot tell which terminal is connected to the human and which terminal is connected to the computer, then the computer has demonstrated some intelligence.

A relatively simple program can carry on a meaningful conversation with a human. One of these is ELIZA which pretends to be a psychotherapist. This program was written as a demonstration of artificial intelligence and can run on a personal computer. The program syntactically takes apart the statements made by the human and rearranges them as a response.

The program has no learning capability and it does not understand a word the human says. The program could be firmware or built-in intelligence, since it does not change.

Learning and Educable Machines

One area of intelligence involves the ability to learn. This ability is generally assessed based on what one has learned. The famous study of dogs by Pavlov shows how learning is a result of repeated cause-and-effect events.

The process of learning from one's mistakes is demonstrated by chess or checkers game programs. All the possible moves for a player are analyzed for the game. Only the legal moves are given to the program and a human plays one side of the game. At the beginning of the game all the pieces are lined up on the board.

When a game is completed, if the machine loses, it removes from its list of possible moves the last move it made. This is how the program learns from its mistakes. It is not able to make the same bad move twice. The program does not know that it has won or lost the game. It only knows that it cannot find a legal move with which to continue the game. It takes a number of losses to become a good player.

Another type of intelligence consists of building one's knowledge on the basis of previous knowledge. The computer game of Animal uses this type of artificial intelligence learning.

The program asks the human player to think of the name of an animal so that the program can try to guess it. If the program cannot guess the name, it asks the human player to type in a question that would distinguish the new animal from the last animal the computer guessed. The program saves all the information given to it and uses this information in future guesses. In a real-time environment the computer needs to interpret sensory inputs in time to use the information for the present task. Artificial intelligence programs work at a batch-level mode. A batch mode of operation means that a device processes the information at its own convenience. Artificial intelligence devices need to become faster and less expensive if they are to give practical senses to environmental control.

As research in artificial intelligence brings new and improved sensor capabilities, this will allow more complicated tasks in HVAC. Artificial intelligence will help enable the system to work with incomplete and uncertain knowledge. It will provide help to select the right equipment or system. It will allow the system to diagnose unexpected problems and work around them.

Fuzzy Logic and Plant Control

One way artificial intelligence may affect HVAC control is by allowing it to handle problems that do not have exact solutions. These require the controller to work with fuzzy logic. Fuzzy logic problems deal with situations that have several gray areas and may have several reasonable solutions. The objective is to find the best of these possible solutions. The use of fuzzy logic can greatly increase the system's functionality and flexibility.

The ideal automated plant takes in raw materials or fuel at one end and outputs finished products at the other end, without the aid of human labor. Humans may provide special talents in problem solving and decision making. They will set up a supply of raw materials and tell the plant what is to be produced and arrange for transport of the finished products.

The automated plant may consist of equipment, robots, and material handlers which are computer controlled. Much of this equipment will have built-in computers. The management within the automated factory will be by expert systems.

Supervisor computers will schedule equipment computers, while a central management computer will schedule the supervisor computers. Human management will control the central management computer. The expert systems will allow the plant to reroute production around defective machines or to start secondary production systems. All the computers and equipment used in the automated plant will have a high level of intelligence.

In its early stages, the automated plant's ability will be to produce a specific product. Later, this will include the ability to produce more than just one product or a closely related family of products.

Understanding Human Languages

Several areas of artificial intelligence deal with understanding human languages. These include voice or speech recognition, and natural language understanding. Voice or speech recognition deals with converting the spoken language of humans into recognizable patterns for the computer. Natural language understanding involves accepting commands in a human language, such as English, and translating these commands into the computer's own machine language. A voice recognition system or a keyboard may be used to get the words into the computer. Then the natural language understanding program finds the meaning for the words.

Computer machine languages are precise while human languages tend to be inexact. The same English sentence conveys different meanings to different persons. The same person may also give different meanings to the same sentence under different conditions. The inexactness of natural languages occurs because the meaning of a sentence is influenced by the context around the sentence.

A very basic natural language understanding system can be built from a 1,000-word, speaker-independent system. A more complete system for natural language understanding would use a 10,000-word vocabulary.

Voice Recognition

Early voice recognition depended on voice prints which are digital recordings of short parts of speech. These prints were analyzed for the length of a word and for the frequencies used in the word. Each voice print acted like a fingerprint and had to be memorized for every person that the computer would receive voice input from.

As the number of words in the vocabulary grows and the number of persons using the system increases, the amount of memory required for storing the voice prints expands and the time needed to search the voice prints for a match goes up. Voice print systems tend to be impractical for handling large vocabularies.

Another type of voice recognition uses phonics to recognize words. This requires storing the known phonemes of speech for a particular language. These phonemes are the smallest units of speech that distinguish one utterance from another.

Each word in the vocabulary is represented by these phonemes. When a voice input is received, the input is broken down into a string of phonemes and the computer makes its best guess at matching the utterances to the phonemes. The computer must also recognize the breaks between each word.

The computer takes the string of received phonemes and compares them to known word phoneme patterns. The computer may ask for the meaning of a new word or for repronounication of the word. If it is a new word, the computer will add the phonemes and definition to its dictionary.

Humans increase the speed of their speech processing by assuming what they expect to hear. This technique occasionally leads to mistakes and the hearer may have to ask for part of the speech to be repeated. All of a person's past experiences are used for reference in speech processing.

Speech understanding involves speech recognition and word understanding. Speech recognition deals with recognizing that a certain series of sounds represents a certain word. Word understanding is natural language understanding and deals with the relationships between different words.

Most people think that they hear and understand about 80% of what is said. The actual percentage can be less than 50%. One factor that causes a low speech recognition percentage is the use of homonyms, words that sound alike, but have different meanings. Following are some typical homonyms:

for	fore	four
their	there	they're
to	too	two

Homonyms can also consist of groups of words like the following phrases:

recognize speech
wreck a nice beach

Also, every person has a unique micro-accent, speaks at different frequencies and varying speeds, and tends to run words together.

Speech understanding starts with hearing sounds, which must then be recognized as words. Finally the words must be combined to produce idea understanding. Hearing is the process of responding to sound waves. This requires some type of specialized sound-wave receptor. A microphone is a transducer that converts sound waves to electrical energy. A microphone is actually a special loudspeaker running backward to detect sounds.

Speech recognition systems are used to recognize spoken words. These may work on one word at a time or on continuous speech. The system may be tuned to a single person's voice, or it may be usable by many different persons. The least expensive systems operate with isolated words and are speaker-dependent.

Speaker-dependent Systems

A speaker-dependent system is a speech recognition system that recognizes only the words or commands spoken by one particular human. A word spoken by a human has a complex waveform that is as unique as a fingerprint. The electronic recordings of human words are called voice prints. A speaker-dependent system is less costly than a speaker-independent system, but it must be trained by having the intended user speak the command words while the system records them. These voice prints are later used by the system for matching the present input.

Most systems digitize the voice print and break it down into its different frequency components. The more information saved about the voice, the more secure the system is from unauthorized use. If the person is talking under stress or in a noisy environment, the system may not recognize the commands given. Most speaker-dependent systems have a limited number of words in their vocabulary. A speaker-dependent system must also be retrained when you switch users.

Speaker-independent Systems

A speaker-independent system is a voice recognition system designed to recognize commands given to it no matter who does the speaking. Since different persons say words at different speeds, frequencies, and inflections, a speaker-independent system must analyze the speech in several steps and then base its identification on the results.

First, it breaks the sounds down into phonemes so it must be able to find the pauses in the words. Then, it looks up the phonemes in its memory and tries to find the words they represent, which can require storing several pronunciations of the words. Then, it compares these words against known commands and selects a command. If a command is not understood, the speech recognition system must request that the command be given again.

For short vocabularies, the voice print technique can be used by a speaker-dependent system. For large vocabularies, the phonic technique takes less memory space and is more efficient.

Isolated-word Recognition Systems

Isolated-word recognition refers to a speech recognition system that only hears one word at a time (Figure 6-1). It processes the word it hears and performs the associated command. If multiple words are spoken to an isolated-word recognition system, it treats them as a single word or command.

Many word recognition systems operate with isolated words. These may be speaker-dependent or speaker-independent. These systems have been used for giving simple commands to a machine or robot.

An isolated-word, speaker-independent voice recognition integrated circuit like the VCP200 can understand the words go, stop, left turn, turn right, and reverse. It can also understand yes/no or on/off.

Olivetti of Italy has worked on a listening and talking machine that can function as a listening typewriter. It acts as an isolated-word, speaker-independent word recognition system with a vocabulary of 10,000 words. When a word is first spoken, the machine selects the best three or four candidates for the actual word and places the likeliest of these in the text. If later words give a better clue to the word just spoken, the machine goes back and changes that word in the text.

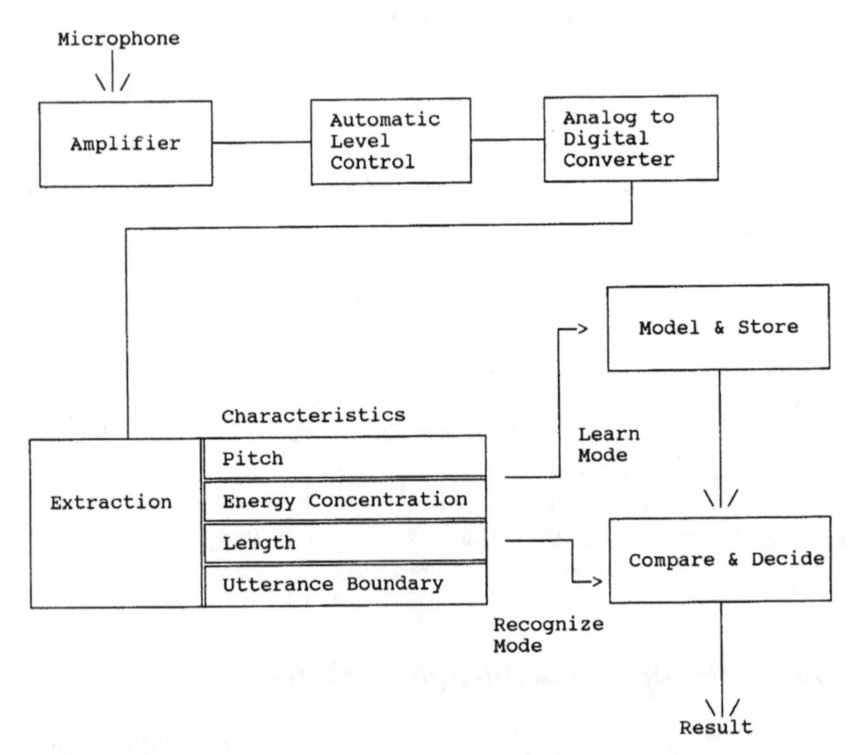

Figure 6-1. Isolated-word recognition system

Continuous-speech Recognition Systems

A continuous-speech recognition system is capable of understanding sentence-length concepts and ideas (Figure 6-2). This type of speech recognition system has to overcome problems that are not encountered by an isolated-word recognition system. A continuous-speech recognition system must work in real time so it must understand what is being said shortly after hearing each word. Continuous-speech recognition

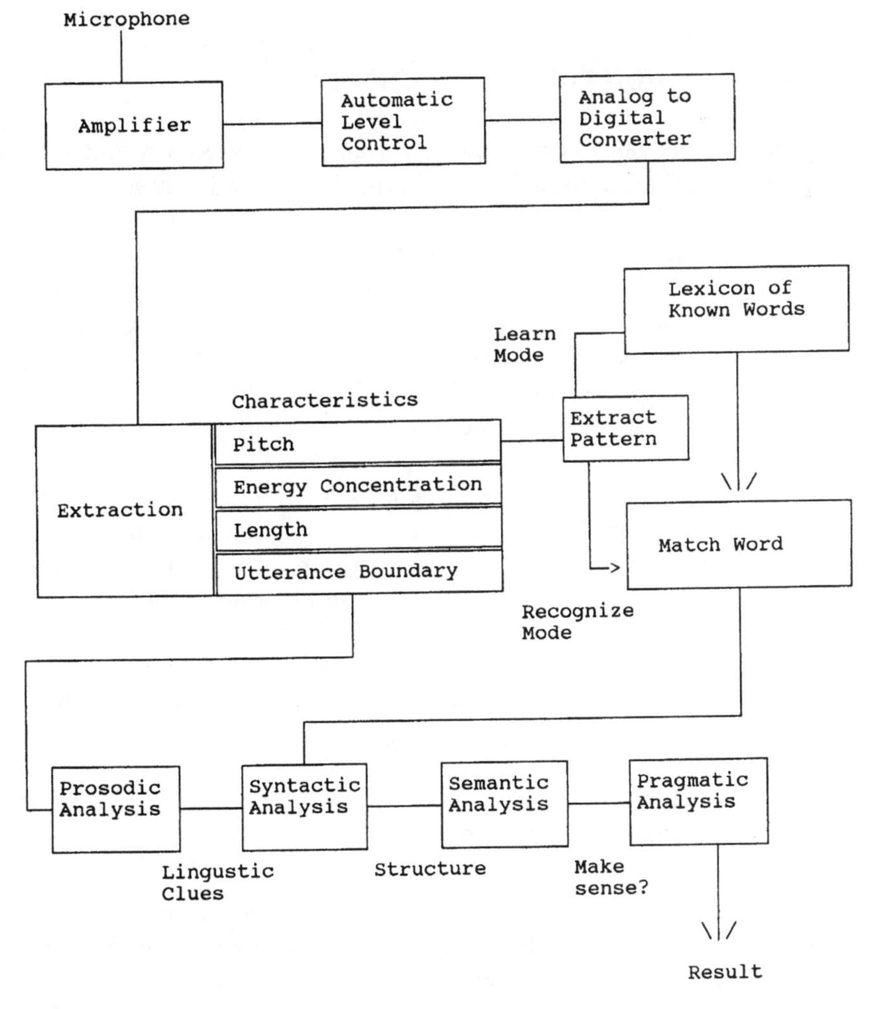

Figure 6-2. Continuous-speed recognition system

systems can use a learning process similar to Morse Code. Learning starts with recognizing the dots, dashes, and spaces. Then, learning proceeds to recognize the characters without having to count the dots and dashes. Next, you learn to recognize simple words and eventually get to predict complete sentences or phrases.

Fast speech recognition requires that the system must be able to recognize common phrases. Since speech recognition systems are used in specific application areas, the language of the particular application like HVAC becomes the heart of the system's common-word vocabulary.

Speech Recognition Techniques

Four approaches or techniques for electronic speech recognition exist. Acoustical techniques depend on the analysis of the speech wave form. Signal analysis can be done with Fourier frequency analysis, principal component analysis and statistical decision making.

The speech perception approach bases speech interpretation on the essential components of speech such as the length of the sounds, the separation of vowel and consonant sounds, feature detection and linguistic categories. Speech production techniques are based on the idea of understanding how speech is produced in order to understand its meaning.

Speech reception techniques duplicate the human auditory reception process in a machine. It is based on observing how the ear receives sound and converts it into electrical impulses, and how the nerves transmit the sound information. The concept is to replicate the process mechanically.

Humans hear the contextual meaning between spoken words and past experiences and these are used for speech interpretation, as well as the physical gestures made by the speaker.

Speech Recognition Problems

Speech recognition systems have some common problems. They need to be able to identify the most probable answer, but the only system that can be expected to find exact matches is a voice print system.

All systems require a means of receiving the sound waves and converting them to electronic signals. This requires some type of microphone and amplifier circuit, as well as some type of automatic level control to make the sounds peak at the same level. Spectrum analysis can be done on the analog signals, as can zero crossing or utterance boundary detection.

The analog signals may also be converted into digital signals and stored in memory. The memory can be scanned and compared to the present input for a voice print match. Spectrum analysis, zero crossing and utterance boundary detection can also be done as digital information stored in memory.

If continuous-word recognition is used, there may be additional tasks of prosopic analysis, syntactic analysis, semantic analysis, and pragmatic analysis. Prosopic analysis tries to find clues to linguistic structures, stressed words and areas of phonetic reliability. Syntactic analysis involves sentence structures and grammatical relations. Semantic analysis checks the interpretation of the words for a meaningful sentence. Pragmatic analysis is used to verify hypothesized word combinations.

Speech Synthesis

Speech synthesis refers to the process by which a machine produces speech. Voice input/output can simplify human/HVAC interactions and make HVAC more user-friendly.

Voice output or speech synthesis is considerably easier to add to a machine than vision processing or voice input. When phonic integrated circuits are used, voice output requires less computing power.

Speech is omnidirectional and understood by most humans. Humans can listen to the speech output of a machine while focusing on some other task. In an emergency, a machine could use voice or speech output to transmit critical information faster than it could by using lights, sirens or displays.

Some of the earliest work on machine speech synthesis is credited to Alexander Graham Bell, in the mid-1800s. Using bellows and other mechanical devices, Bell tried to build a copy of the human speech mechanism. He was able to simulate a few words and this work contributed to his invention of the telephone.

Three methods can be used to produce machine speech output:

- prerecorded messages produced by human voices,
- digitized words produced by human voices and
- phonic integrated circuits.

Early electronic speech units used tape recordings of a human voice. The human voice is an analog form of energy made up of a mixture of tones. If prerecorded messages are stored on a single magnetic tape, the tape will have to be searched each time for the correct message. This time delay could be a problem in an emergency.

Samples of prerecorded human voice output are still used in some toys. A miniature record is used. Digital audio recordings can use a compact disk for high-quality sound, and a large amount of memory can be accessed in less than 1 second.

A system that uses a digitized human voice is able to store the sounds in computer memory, like any other digital information. The sounds made by a human being are converted and stored as digital information.

A National Semiconductor digitized human voice system called Digitalker used from 3 to 5 special ICs along with several support ICs for address selection and audio amplification. The main IC was a MM54104 microprocessor-based controller. The ROM ICs held a total of 274 words. Any message made up of these words can be spoken by the Digitalker. Special ROMs are used to satisfy a specific application.

The Digitalker interfaces to a personal computer through a parallel printer port. Programming the Digitalker involves sending it a series of hexadecimal addresses of words that are to be spoken. The addresses for the proper pauses between words must also be included.

An inexpensive way of producing speech output for machines is to use phonic integrated circuits. A phonic integrated circuit uses phonemes which are the smallest distinct units of speech to produce speech.

A typical phonic integrated circuit may have 64 phonemes, which includes 5 different lengths of pauses. The other 59 phonemes are the sounds such as "OY" in boy and "AY" in sky. See Table 6-1.

A processor and an audio amplifier forms a working system with these ICs. A sound is produced by giving the processor the address of the desired phoneme. By giving the processor a series of addresses, you can produce words and sentences. To produce the word MAY, for example, you could address an MM, as in milk, and then an EY, as in beige.

Table 6-1. Typical Integrated Circuit Phonemes

Hexadecimal Address	*Phonemes*	*Meaning or Sound*
00	PA1	Pause 10 MS
01	PA2	Pause 30 MS
02	PA3	Pause 50 MS
03	PA4	Pause 200 MS
3A	OR	Store
3B	AR	Alarm
3C	YR	Clear

Producing the word six is more complex. It requires:

- two SS sounds as in vest,
- two IH sounds as in sit,
- a pause of 50 milliseconds,
- KK sound as in sky and
- SS sound as in vest.

Using phonics requires some processing time.

Most phonic integrated circuits cannot remember a series of commands. They work on only one command at a time. Additional circuits are required to get the integrated circuit to produce more than isolated sounds.

The CTS256A-AL2 text-to-speech controller IC handles many of the support functions for the SP0256-AL2 Speech synthesizer IC. It accepts ASCII characters as input and outputs the phoneme codes to the SP0256-AL2. The CTS256SA-AL2 furnishes a buffer for the ASCII characters, and it can forward the characters for pronunciation as single letters, single words, or groups of words separated by appropriate punctuation marks.

The sounds in an integrated circuit are always pronounced the same way each time. While the base frequency of the voice can be varied, it is still not a pleasant voice. The phonemes do not represent regional speech differences and do not handle accents well.

These phonemes are made for American English and if the speech system were used to pronounce Spanish or French, it would be hard to recognize what is being said. Each spoken language has its own combination of phonemes. Programming the SP0256-AL2 involves sending a series of addresses to the IC. These addresses must include phonemes for each word, including the pauses within each word and the pauses between words. User-defined speeches consist of machine language instruction placed around the phoneme addresses. Speech or voice synthesizers are available for personal computers. Table 6-2 shows a list of stored words for digitized voice systems.

Table 6-2. Typical Stored Words for Digitized Voice System

Word	*Hexadecimal Address*
Emergency	23
End	24
Enter	25
Entry	26
Evacuate	28
Exit	29
Fail	2A
Failure	2B

Expert Systems

Expert systems mimic the tasks usually carried out by experts. The expertise of these systems is confined to a well-defined area. Unlike humans, these systems lack common sense and cannot extrapolate their knowledge beyond their narrow domain of expertise. Expert systems act like a storehouse of knowledge with ways of retrieving this knowledge to solve a particular problem (Figure 6-3).

An expert system is a computer program that has some information that helps it to make a decision. One simple example is a troubleshooting table given in a product manual. A set of procedures for

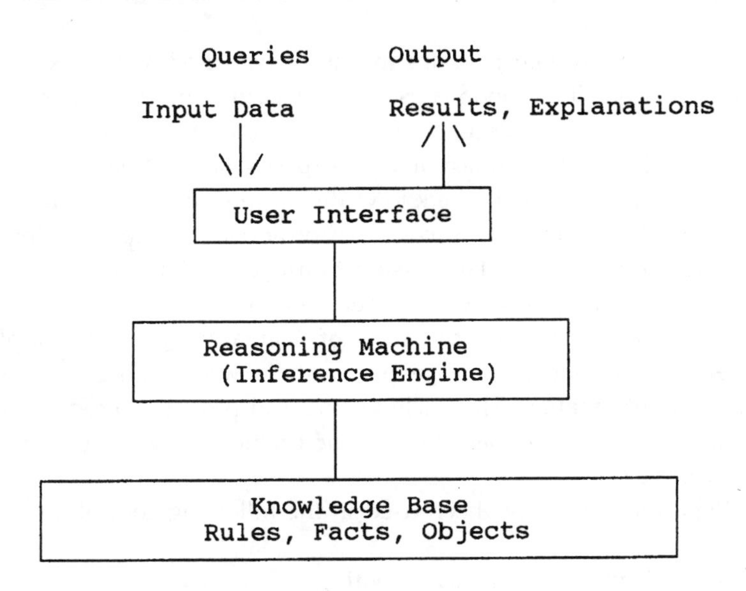

Figure 6-3. Expert system structure

troubleshooting a computer system is also known as a diagnostic decision logic table (DDLT).

Robotic systems are using expert systems to decide which robot is best for performing a particular task and to help troubleshoot broken robots. Expert systems are also used to schedule factory production and maintenance.

An expert system can have four parts:

- inference engine,
- knowledge base,
- database and
- natural language interface.

The inference engine acts as a rule interpreter which matches facts and rule conditions. It is the controller of the system. The knowledge base contains the rules while the database collects the facts given to the system by the user. These facts are used by the inference engine in deciding what rules to apply to the problem. The natural language interface allows the user to communicate with the expert system in a natural language such as English or French.

The expert system operates in one of two modes. The expert or knowledge acquisition mode is used to enter rules into the system. The user or consultation mode is used to solve the problem (Figure 6-4).

The game of Animal acts like an expert system. When it asks for questions, it is operating in the knowledge acquisition mode. It uses the answers to these questions to construct or add to its expert system of animal classification. Similar classification expert systems are used in medicine, publishing, geology, and construction.

The expert system for the game of Animal works with complete and certain information. Other expert systems work with uncertain or incomplete information. This includes medical programs used to diagnose and prescribe treatment for specific problems like bacterial infections of the blood.

Expert systems have been used in the following control areas:

- fault diagnosis and troubleshooting,
- quality control,
- control system synthesis,
- statistical control and
- alarm management.

Artificial neural networks (ANN) can be used as a tool for examining data and determining relationships. They are useful for capturing relationships among measurements and modeling. Compared to the technology of expert systems, the technology of neural networks is newer.

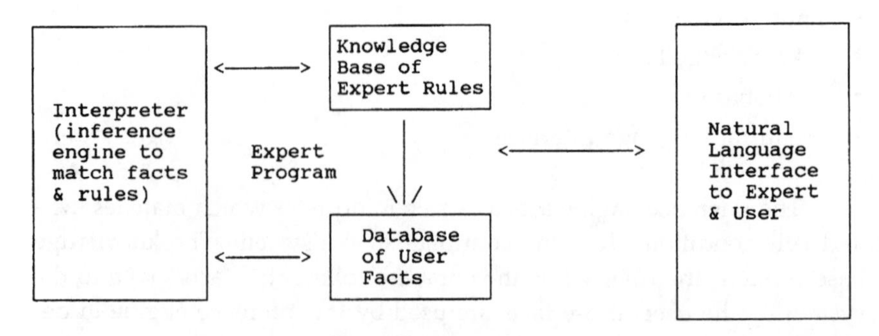

Figure 6-4. One type of expert system operates in an expert mode while learning new facts and in a user mode while using its knowledge base.

Structure of Expert Systems

Expert systems encode the relevant knowledge about the problem domain in a form which is suitable for manipulation. The structure of the expert system separates the knowledge from the mechanism for manipulating the knowledge.

It is this separation of the knowledge base from the reasoning process that makes expert systems different from most other software packages. The type of knowledge in the knowledge base depends on the problem. Knowledge is stored in the following ways:

- Acts — As temperature goes up, fuel use increases.
- Rules — If space temperature is high, then the HVAC system is faulty.
- Data frames — Data structures for storing information in a hierarchical format.
- Objects — These perform specific tasks such as an energy balance.

Objects may be coded in other programming languages and both the knowledge and the data can be either qualitative or quantitative since expert systems have the ability to deal with qualitative data and knowledge.

The problem-solving strategy of the expert system is contained in the inference engine. The rules in the knowledge base are applied in a certain sequence until a conclusion is reached. This process is called rule chaining.

The user interface provides a mechanism for the expert to enter knowledge and the user to enter data for a specific problem. Most expert systems are programmed using expert systems shells.

An expert system shell contains all of the structure, but its knowledge base is empty. By programming the shell to contain knowledge about a domain, an expert system can be built up.

Real-time expert systems involve the time element in the reasoning process and must be integrated with sensor and data communications with the operator (Figure 6-5). Examples of real-time expert systems include G2 from Gensym and Expert 90 from Bailey Controls.

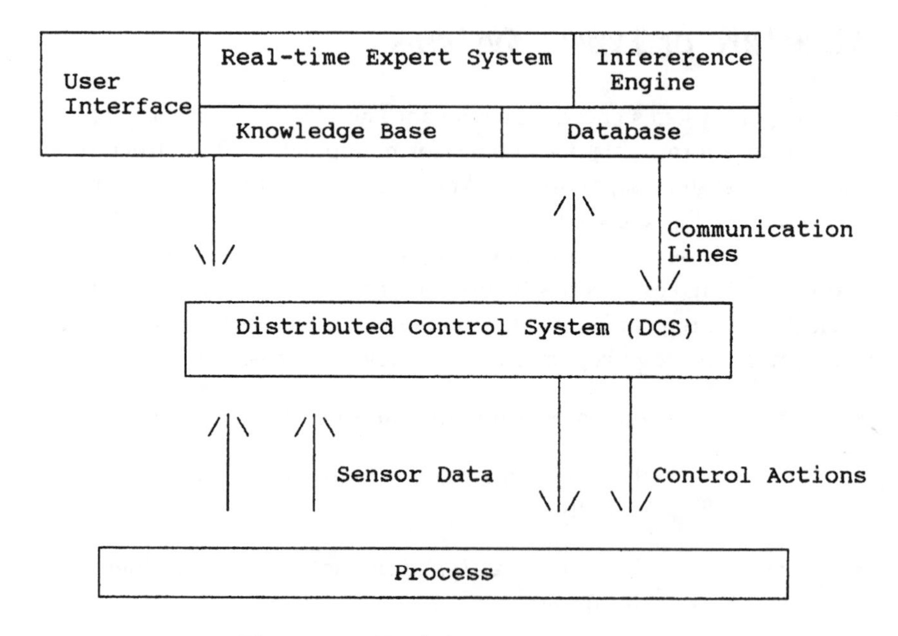

Figure 6-5. Real-time expert system

Knowledge acquisition is the main problem in the implementation of expert systems. One technique is to use knowledge engineers to interview experts and then incorporate the knowledge into an expert system. Another technique is to have the expert enter the knowledge into the expert system shell. This technique has been more popular.

Most shells are designed to allow the knowledge base construction to be automated. Some shells generate the knowledge base from examples.

Expert Systems in Control

Expert system applications include both control system design and real-time control. In expert systems for design, the goal is to reduce the design time and to produce optional designs. The expert system gathers the knowledge, organizes it and codes it into the program. The user interacts with the program in an interview mode, supplying relevant data on a specific problem. The expert system will suggest an appropriate control configuration and several options. The advantage of such an expert system is that it can be reproduced, distributed and the expertise becomes easily available.

The expert system can be looked at as a way of capturing the expert knowledge in specialized areas. However, the knowledge contained in an expert system is never complete and the output of an expert system must be validated.

The expert system can capture diagnostic knowledge and code it in a software package that could then be used by operators to diagnose routine problems. A convenient way to represent the knowledge is in the form of a decision tree with each node in the tree representing a decision to be made regarding the possible cause of the problem.

Expert systems can serve as an additional tool for the inexperienced operator in diagnosis of faults. The expert system can diagnose the cause of the fault and suggest actions to correct the situation.

The user interface is critical as well as the type of queries made by the expert system during the diagnostic session. Other critical parts include the language and terminology used, the use of graphics and the response time of the system to entries made by the user.

Statistical Control

On-line statistical control is a good tool for monitoring if a process is operating normally. Many DCS systems generate quality control charts automatically but the operator has to decide the action to take if some abnormality is detected. This can be done using some type of Pareto analysis.

On-line expert systems are sometimes called DCS-embedded expert systems and they can do this diagnosis automatically. The DCS can feed the current state of the process to the expert system at pre-specified intervals of time. Messages on the operator console provide alarms that not only point to the problem but also provide advice on possible causes and solutions.

On-line expert systems may acquire their knowledge base automatically. This falls under the domain of machine learning. In a large HVAC system there may be a large amount of data that is generated from sensors. This data must be reduced to a form that can be used by an expert system.

Future Trends

Expert systems open a new dimension in the way computers can be used for HVAC. Expert systems technology is being integrated into the architecture of many distributed control systems and networked programmable controller systems. Self-tuning PID controllers use expert system technology. A boiler might incorporate a troubleshooting and diagnostic expert system as part of the control system for the boiler.

Expert system applications include sensor testing and validation, control system design validation, performance evaluation, diagnostics, on-line tuning and statistical control and analysis.

Neural Networks

Neural networks such as expert systems, fuzzy logic, robotics, natural language processing, and machine vision, are parts of the overall technology umbrella known as artificial intelligence (AI). Neural networks attempt to mimic the structures and process of biological neural systems. They provide a powerful analysis technique for complex processing of large amounts of input/output information. They have the ability to generalize or form concepts.

Artificial neural networks (ANNs) are a useful information abstraction in modeling intelligence in control systems. Neural networks may be thought of as a functional mapping of inputs to outputs using an interconnected network of nodes. Weights are given to the node interconnections to achieve the desired mapping.

Artificial neural networks are a cognitive information processing structure based upon the models of brain function. It is a highly parallel dynamic system that processes information using the state response to inputs.

The advantage of neural networks is in their ability to learn arbitrary function mapping with little or no prior knowledge about the function itself. They provide the capability to do black-box modeling of a process given only the Input/Output data.

These networks have proved to be robust, resilient and capable of adaptive learning. The disadvantage is that the knowledge is implied by the network connection weights.

Neural networks have been used as virtual or soft sensors to infer quality from sensor data. DuPont has used them to predict the composition of chemical mixtures.

The inputs to the network are measurements of the mixture. Training the network to determine the network connection weights involves data from mixtures whose composition is known. Numerical algorithms to train the network are used from commercial software packages. DuPont has also used neural networking as a virtual sensor to predict periodically measured concentrations.

Historical data on temperature, pressure, and concentration were used to train a network to predict the concentration from temperature and pressure measurements. This trained network can then be used to provide continuous, on-line prediction of concentration.

Neural networks are a tool to solve many complex problems. They may be embedded in databases or expert system applications or act as preprocessors or postprocessors to other systems.

Neural networks have the ability to adapt, generalize, and extrapolate results. However, they cannot optimize and need lots of data. They can be unpredictable in untrained areas and are not well understood or widely accepted.

Neutral Network Growth

The basic concepts of the neural network have been known since the 1940s. The theoretical foundations were established in the 1960s and 1970s. Research of the 1980s-1990s provided the first applications.

The researchers included neuroscientists, cognitive psychologists, physicists, computer scientists, mathematicians, and engineers. The increasing power and inexpensive cost of computing allowed the development and deployment of systems for industrial applications.

Neural networks obtain their name and some of their associated terminology from biological systems (Figure 6-6). Neural networks are built of neurons which are also called nodes or processing elements.

These nodes are usually arranged in layers or slabs and are often connected to nodes in other layers. A layer is a set of nodes with weights that are actively manipulated. These layers serve as a buffer between the inputs or outputs or other layers.

A slab is a set of nodes that may be different in their internal specifications or connectivity but which share the same layer. A single layer

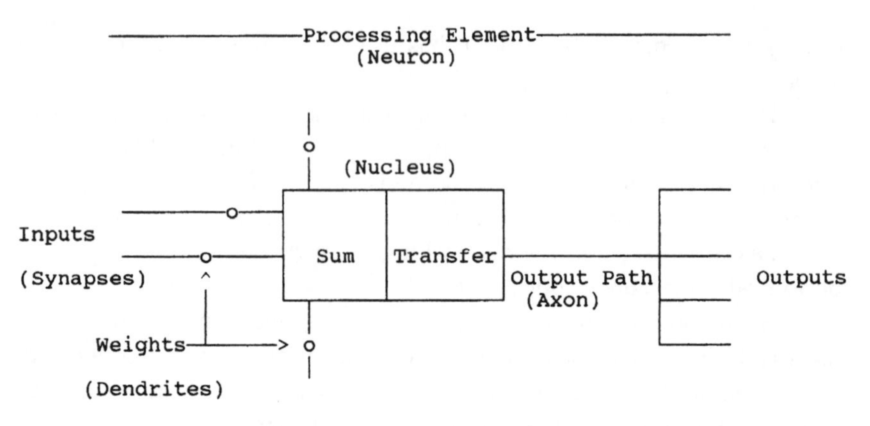

Figure 6-6. Neural net processing analogy

can consist of multiple slabs.

Each node processes the input it receives, uses these connections and provides a continuous output to other processing elements using its outgoing connections. Like biological systems, the strengths of these connections may change.

The different ways the processing elements in a network are connected, how the processing elements process their information, and how the connection strengths are changed make up the neural network. There are more than 100 different types of neural networks and this number is still growing.

Linear programs and other older modeling techniques require that an algorithm describe how the system should proceed. Data is fetched when needed for execution of the instructions and the results stored. These operations are performed in a deterministic and sequential manner.

Neural networks consist of heavily interconnected processing elements that do not store information. The knowledge that is stored depends on the way the processing elements are connected and by the importance of each connection, known as its weight.

Rather than being programmed, neural networks are trained. The network adjusts itself based on a specified learning rule that corrects the response to a desired response. Many representative samples are needed by a neural network in order for it to learn its rules for processing knowledge.

Neural Network Operation

A neural network processing element has many input paths which are individually multiplied by a weight and then summed. A nonlinear transfer function known as a squashing function is applied to the result to calculate each processing element's output.

The transfer function must provide both nonlinearity and stability to the network. Typical transfer functions include the sigmoid transfer function which is most often used in back-propagation networks and TanH which is a bipolar version of the sigmoid.

The output value of the transfer function is usually passed directly to the output path of the processing element. The output path is then connected to input paths of other processing elements through connection weights. The weights and connections form the memory or knowledge of the neural net. Since each connection has a corresponding weight, the signals on the input lines to a processing element are modified by these weights prior to being summed. Thus, the summation function is a weighted summation.

A neural network consists of many processing elements joined together. A typical network consists of a sequence of layers with connections between successive layers. A minimum of two layers is required. These are the input buffer where data is presented and the output layer where the results are held. Many networks also use intermediate layers called hidden layers.

Applications for neural networks need an abundance of historical data or examples with data dependence on several interacting parameters. Back-propagation neural networks have been used in chemical process control to predict boiling points.

The type and number of processing elements, the number of slabs, the number of layers, the connectivity of the layers, the transfer function, the learning algorithm or rule are all parameters of the network. Other parameters include the learning threshold and learning coefficients and learning schedule.

The learning schedule is a breakpoint table that allows the learning coefficient to decay after a number of learning passes. The higher the learning coefficient, the faster the learning. However, the higher the learning coefficient, the slower the convergence. Therefore, if the learning coefficient can be reduced as training proceeds, high learning and fast convergence can take place. The momentum factor acts as a low-pass filter on the weight differences that allows faster learning on low learning coefficients. Other features to expedite convergence include algorithms for random weight generation, introduction of noise as the network learns, and algorithms to jog the weights randomly.

Hidden layers act as layers of abstraction and help the network generalize and memorize. Most control applications require only one hidden layer.

One way of determining whether multiple layers or multiple slabs improve the performance of a network is experimentation. Determining the number of nodes in a hidden layer is also an experimental exercise.

More complex relationships require more processing elements in the hidden layer. Too few nodes in the hidden layer prevents the network from properly mapping inputs to outputs. Too many nodes promotes memorization and inhibits generalization.

Memorization occurs when the patterns presented to the network are reproduced exactly without extracting any salient features. The network is then unable to process new patterns correctly because it has not discovered the proper relationships.

Preparing the data includes transforming inputs into the proper form such as ratios or classes and data types. How the data is represented and translated plays a role in the network's ability to understand during training. Data may be continuous, digital, time-oriented or static. Data can be naturally grouped, may be represented as actual amounts or changes in amounts, or may be evenly or unevenly distributed over the entire range. When naturally occurring groups appear, binary categories are often the best method for making correlations.

All data needs to be normalized in order for the transform function to operate. Data may be scaled between minimum and maximum ranges or be set between a predefined range.

Often the network is trained on one set of data (training data) and verified with a different set of data (recall data). Once a network is trained on a set of data, it is used to predict results based upon new sets of gathered data.

Statistical tools can help determine how the network produces the correct outputs. Small weights indicate that the processing elements are not influencing the outcome of the network. These nodes can then be eliminated and the network retrained. Large weights can indicate too much dependence upon a particular input, indicating some degree of memorization.

Fuzzy Control

Fuzzy controllers have rules that are used to calculate the control action. Each rule is an expression about the control action to be taken in response to a given set of conditions. These rules have the following general format:

IF (CONDITION) THEN ACTION

The condition may include AND/OR operations. The fuzzy controller uses the inference process, (Figure 6-7), which has several rule processes and produces a single logical sum. Each rule process has its conditions (antecedent) and a conclusion (consequence). The conclusion is reached when the conditions are satisfied. The conclusions are then sent through a logical summer.

The defuzzification operation then unifies the results of the rule process and produces a final output that can be interfaced to most types of output devices.

The problem of energy use in HVAC can be used to illustrate the operation of a fuzzy controller. A set of instinctive judgments is trans-

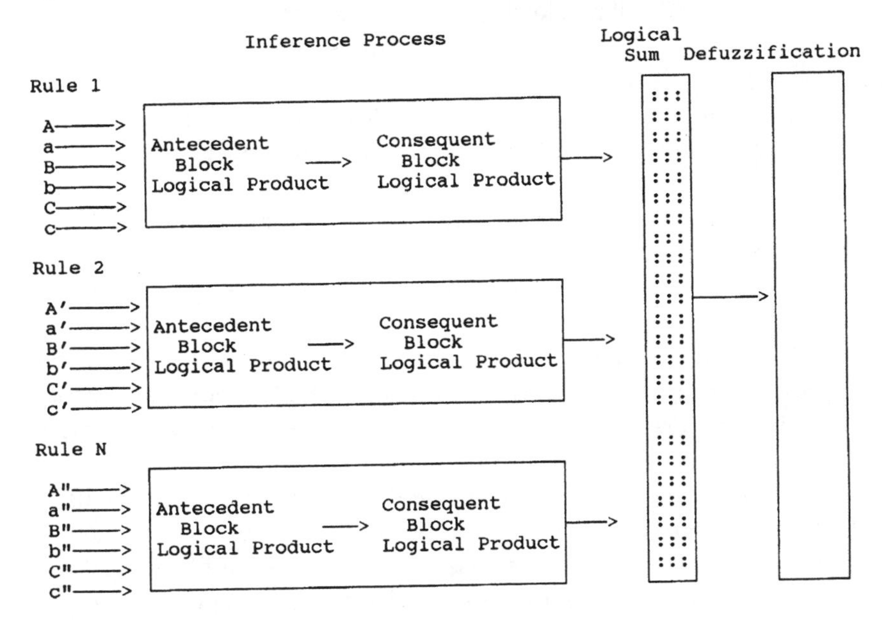

Figure 6-7. Operation of fuzzy control

planted into the machine, without using complex equations.

Temperature balancing can be done with the IF...THEN function in fuzzy control (See Table 6-3).

Table 6-3. Rules for Fuzzy Temperature Control

	Antecedent Block	*Consequent Block*
Rule 1	If temperature is moderately low and almost still	Then reduce fuel moderately but quickly
Rule 2	If temperature is moderately low and falling slowly	Then increase fuel moderately and a little quickly
Rule 3	If temperature is moderately low and rising slowly	Then do not change fuel flow much
Rule 4	If temperature is moderately high and almost still	Then decrease fuel moderately but quickly
Rule 5	If temperature is moderately high and falling slowly	Then decrease fuel moderately and a little quickly
Rule 6	If temperature is moderately high and rising slowly	Then do not change fuel flow much
Rule 7	If temperature is at setpoint and almost still	Then do not change fuel flow much

The antecedent blocks (IF... phrases describe the states that the space can be in, and the consequent blocks (THEN... phrases) describe how the temperature should change in response to these states (See Table 6-4).

Fuzzy expressions such as moderate, almost, and a little are used. If the condition is true, the measure of fulfillment of the statement is equal to 1.0, if the condition is false, the measure of fulfillment is 0.0. A value between 0.0 and 1.0 is possible for anything between true and false.

Table 6-4. Antecedent and Consequent Blocks for Fuzzy Temperature Control

	Antecedent Blocks Antecedent Membership Functions			*Logical Product Minimum*		*Consequent Blocks*
Rule 1	0.7 PM	0.7 AND	ZE 0.8	0.8 = 0.7	THEN	PM 0.7
Rule 2	0.3 PS	0.3 AND	PS 0.2	0.2 = 0.2	THEN	PS 0.2
Rule 3	0.3 PS	0.3 AND	NS	0.0 = 0.0	THEN	ZE
Rule 4	NM	0.0 AND	ZE 0.8	0.8 = 0.0	THEN	NM
Rule 5	NS	0.0 AND	NS	0.0 = 0.0	THEN	NS
Rule 6	NS	0.0 AND	PS 0.2	0.2 = 0.0	THEN	ZE
Rule 7	ZE	0.0 AND	ZR 0.8	0.0 = 0.0	THEN	ZE

Key:
NG Negate
NL Negative Large
NM Negative Medium
NS Negative Small
PM Positive Medium
PS Positive Small
ZE Zero

There are seven rules listed, but fewer can be used. When fewer rules are used, the function becomes simpler, but control is not as smooth.

A membership curve is used for logical summing. This can take a triangle shape, S, Z, bell shape or trapezoidal. The output depends on how the individual curves overlap and it determines the type and degree of action required.

Rules can be written to include:

- the difference of the temperature from the setpoint,
- the rate at which the temperature is changing.

The logical product of a fuzzy logic function in response to a specific set of inputs always occurs at a point in time. These calculations are similar to the scan cycle of a PLC.

The logical sum combines the results of the rules. The fuzzy device does not make a quick decision. The logical sum may suggest that the controlled parameter should be decreased, but whether this should be done quickly or slowly is left fuzzy unless the result is TRUE or FALSE. This is sometimes called the conclusion of the fuzzy interference.

Defuzzification

The defuzzification operation calculates the center of gravity of the fuzzy interference. This value becomes the output. When the output is a single value determining the required change, the defuzzification is known as the Mamdani method and is the most common technique used.

For multivariable control, the final control action is calculated from the weighted individual membership values of each active rule.

Based on the rules programmed into the fuzzy controller, it will output the most valid value corresponding to the variable input conditions. With fuzzy control, the steps are performed continuously, while in information processing, such procedures are executed only when the input data varies. In one fuzzy control system application the variability in product properties was reduced by about 30%.

Temperature Overshoot Control

A fuzzy logic controller can reduce the temperature overshoot that occurs during heat-up. The ideal control system might start with a low setpoint and then gradually adjust it toward the correct setpoint while watching the control performance. Fuzzy logic can be used for this type of control (Figure 6-8).

Autotuning is used to obtain the PID constants and the process characteristics such as the dead time and time constant. The controller manipulates the heat input based on the deviation from the setpoint. It automatically changes the internal setpoint to a lower value when the anticipatory logic predicts an overshoot. The controller continues to monitor

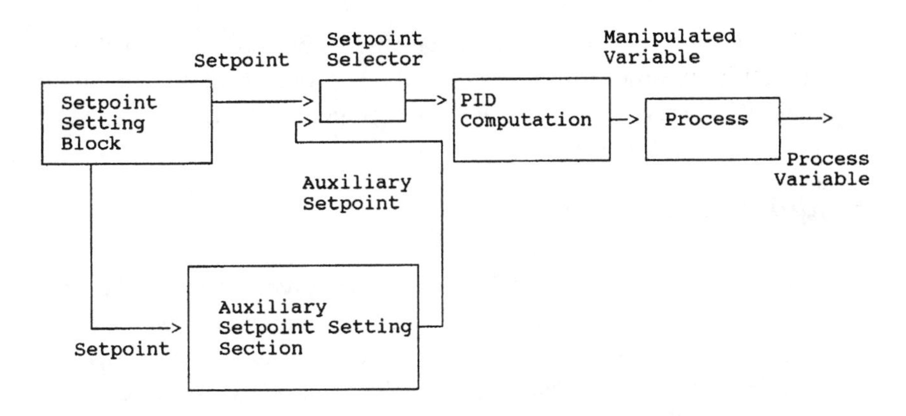

Figure 6-8. PID and fuzzy control

the deviation and returns the setpoint slowly to the correct value, while keeping it within a range where there is no risk to overshoot.

Fuzzy Feedforward Control

Further improvements in product consistency are possible when a process parameter is used as an advanced warning signal. The feedforward input serves to change the fuel flow long before output changes result in environment or product. The feedforward changes are not exact, and their influence is adjustable as in feedback control. The method of calculation of the final feedforward control action is similar to the calculation of feedback control action. In one application the fuzzy feedforward control resulted in a 40% reduction in the standard deviation of the target property.

Fuzzy Color Sensing

Color sensing can be done with fuzzy logic control. A sensor head with light emitting diodes for red, green and blue light is used. The sensor head also has circuitry to measure the light reflected by the object.

A number of noncontact color profile measurements are taken to compute the average color profile. These profiles are compared to a prerecorded standard and if the deviation between the two values exceeds the programmed acceptable tolerance, the output signals a rejection.

Fuzzy logic rules are used to train the sensor to distinguish the designated color and also to compute the sensor output signal. Color acceptability corresponds to the logic set range of 0 to 1.

Model-based Control

The basic concept of model-based control involves a model of the process which is used to calculate a value for the manipulated variable (m) (Fig. 6-9). This value should make the controller variable (x) behave in some desired way. If this is satisfied, then this value of m is utilized.

The controller does not use PID components and usually there is only one tuning parameter, the speed of the controlled variable as it moves to the setpoint. Initializing the controller with a validated model is a critical implementation step.

The inverse model is used. In standard modeling, the input is set and the response of the output is observed. The inverse model determines what input causes a predetermined output response. (Figure 6-10)

If the model is exact, an open loop structure could be used. But, most controller models are not exact and they require some feedback correction. The usual approach is to monitor the difference between the model and the process output and adjust a controller feedback signal. The feedback signal is either a bias to the setpoint or a model coefficient.

Modeling Techniques

Three distinct modeling approaches are used:

- transfer function,
- time series,
- nonlinear phenomenological.

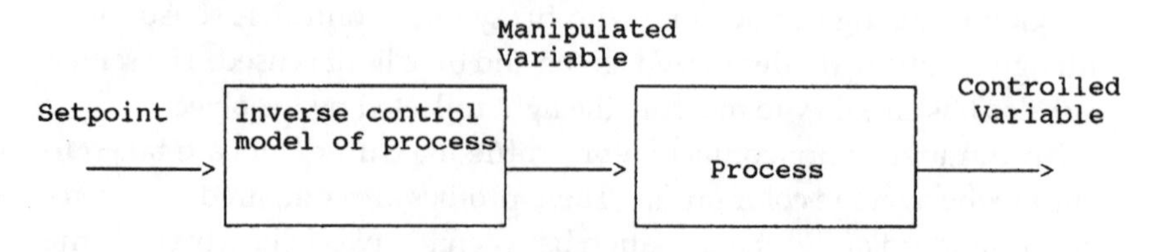

Figure 6-9. Model-based control

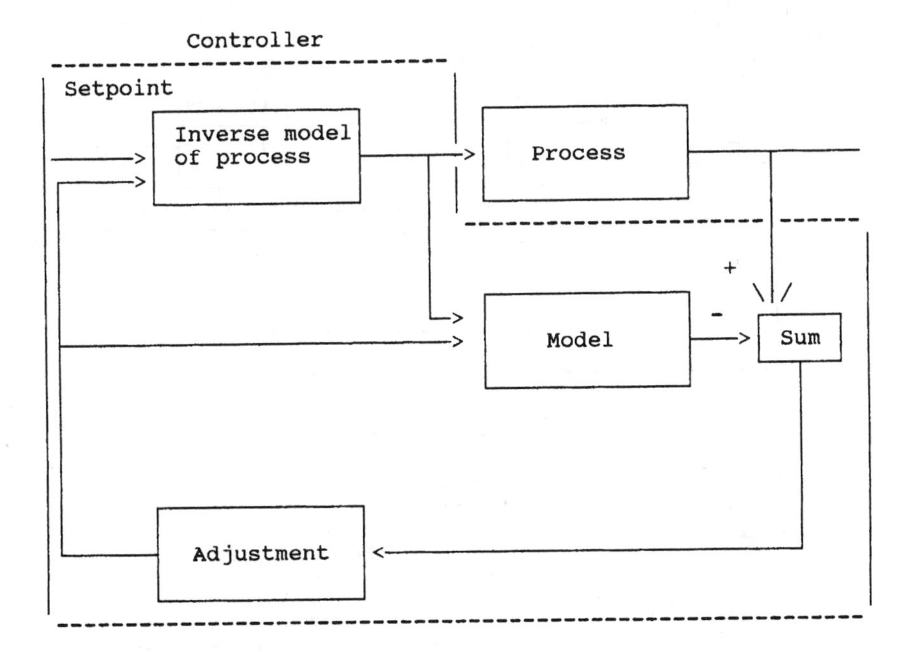

Figure 6-10. Feedback with inverse, model and adjustment functions

Transfer function models use an open-loop Laplace transform description of the process response to a step input. This is a common control modeling approach but is limited to linear and simplistic dynamic modeling.

Time series models involve the open-loop response of the process to a vector of impulses. These are empirically determined and consist of about 30 elements. More precision is possible but matrix/vector algebra is required. This is the most common modeling approach used for model-based control.

Nonlinear phenomenological models are design-based simulators for nonlinear or nonstationary processes. Their control intelligence comes with modeling and computational complexity.

Internal Mode Control

This is a type of modeling that uses open-loop, step-response Laplace transfer functions. The basic structure is shown in Figure 6-11. A simple first-order-plus-dead-time representation is shown in Figure 6-12.

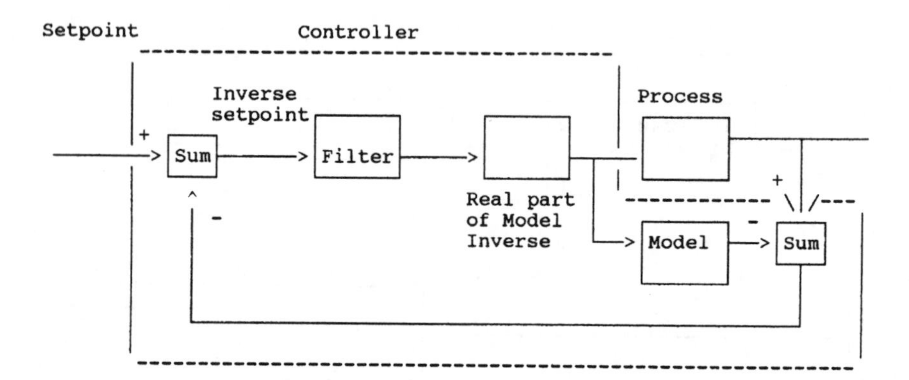

Figure 6-11. Internal mode control

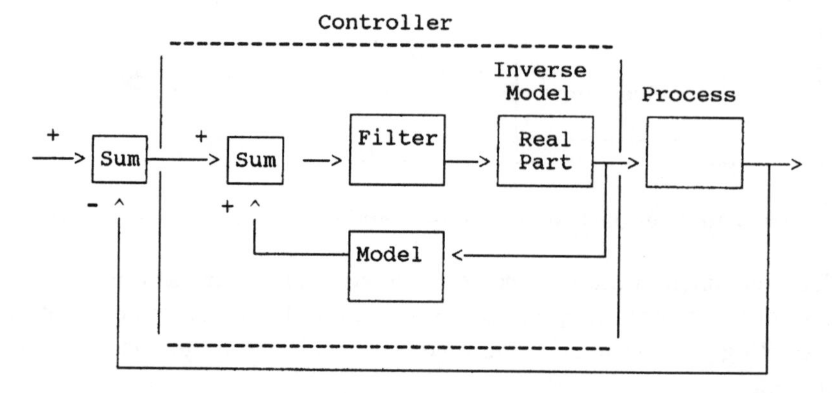

Figure 6-12. Internal mode control with first-order plus dead time controller

Setpoint biasing is the feedback adjustment method. To eliminate spikes in from measurement noise and setpoint changes, the biased setpoint is filtered.

The function blocks are leads, lags, delays, summations, and gains, and they can be configured by most distributed control systems (DCS), single-loop controller or programmable logic controller (PLC). The usual tuning parameter is the filter time constant. Lower values of the time constant make the controller more aggressive and higher values make it gentler.

In internal modeling control the model is linear and stationary. If the process gain or time constants change, the model-based calculations become either too aggressive or sluggish and retuning may be required.

This is similar to PID control where changes in process dynamics also require retuning.

Model Predictive Control

This type of modeling is often called time series or convolution modeling. It represents the open-loop process response as a vector. A single, unity input step change in the manipulated variable is used to affect the output. The input behavior is represented as a sequence of steps and the output as the cumulative effects of those steps.

The desired behavior for the output may be a path of return to setpoint. It can be expressed as a vector and the necessary sequence of manipulated variable values can be calculated from the model inverse.

If the process has dead time, elements of the process model vector become zero. Another problem is that the residual values of the process model vector will affect the output. This can result in a steady-state offset. The methods for handling these problems involve using a filter to remove measurement noise from the model mismatch signal.

The filtered value is used as a reference bias which eliminates the zeros and offset. These controllers require that the process fit a linear and stationary model.

Process Model-based Control

Models can be developed from material and energy balances and thermodynamic relations. These models can be used for economic optimization. Simple, reduced, or tendency models can be used. A model may be represented as

$$\frac{dx}{dt} = f(x,m,p)$$

where

x = output
m = manipulated variable
p = an adjustable process parameter

The control objective is to have the process output move to the setpoint (X_{sp}) within some specific time t1. This specifies above derivative and the control law becomes the model inverse which determines m from the following:

$$f(x,m,p) - \frac{x_{sp.} - x}{t1} = 0$$

Since x is measured, p is set, and x_{sp} and t1 are user-defined, the equation can be solved for m. The tuning parameter is t1. These equations are often nonlinear, and the solution may require iterative techniques such as Newton's.

A steady-state offset occurs if the model is not perfect. Feedback adjustment includes biasing the setpoint and adjusting a process parameter. Adjustable parameters represent a real process mechanism, not a hypothetical disturbance. These adjustable parameters can include heat exchanger fouling factors and ambient heat losses.

Multivariables

There are often measurable loads which affect the output in HVAC as well as other inputs in addition to the manipulated variable. Their effects on the output can be modeled to achieve feedforward action. If there are more manipulatable inputs than controlled outputs, a decision mechanism is used to associate all controlled variables with at least one manipulated variable.

The new information technology in use today enables control systems to function on nonproprietary networks. The controllers often are PC platforms. Typically, the user interfaces are the familiar, commercially available operating systems.

Data and Information

All of this ability to communicate large quantities of data in a short period of time means data overload can become a likely problem. This is the task of control PCs' software interface. It must create useful operator information from the plant data.

One area where this software capability to turn data into information is valve diagnostics. Control valves make many small movements each day to control parameters such as pressure, temperature, or flow rate to an optimum level for efficient plant operation.

Diagnostic tools can measure valve and valve actuator characteristics to determine whether or not such control valves are functioning properly or require maintenance. Analyzing the valve actuator pressure versus valve travel can provide information on friction, actuator spring compression and galling. This information can be important in reducing downtime and maintaining optimum system performance.

Fieldbus communication protocols can provide the two-way communications needed. Using this technology, a smart valve positioner can send information back to the controller concerning its actual final position using the same wires as the outgoing signal.

The sensors throughout the plant also provide an opportunity for significant operational improvements. Advanced calibration monitoring software uses mathematical techniques to predict an expected value such as temperature or pressure. The software computes the statistical variation from the predicted value. This software keeps track of the amount of time each sensor is beyond the setpoint limit and generates a report that flags those devices that are in need of maintenance.

A new generation of valve controllers, called digital valve controllers, provides valve diagnostic capabilities. This is part of the evolution in distributed controls. Low-power microprocessors allow these digital valve controllers to operate on the power from the standard 4-20 mA instrument wiring. The HART communications protocol superimposes a digital signal on top of the 4-20 mA analog signal. This provides the control loop along with two-way communications with the valve controller.

PC and Windows-based software can communicate with the controller to obtain the valve diagnostic variables. The data can be plotted with the software to determine the condition of the valve.

Neural Network Modeling

Besides linking the plant operations with the control of maintenance activities, an additional part of enterprise management is system modeling and the resulting analysis.

Recently, some companies have turned to neural network models, which are useful for dealing with complex situations. As plant operators gain confidence in this relatively new technology, they are even beginning to use it for real-time plant control.

Some industries have used predictive control technology, referred to as model predictive control (MPC), for over 20 years using linear modeling techniques. The increasing complexity of some plants has led to the development of nonlinear models.

A hybrid alternative to these complex, nonlinear models, is a linear model combined with the learning capability of neural networks that helps the system cope with real-world nonlinearities.

A neural network controller does not know how to treat data that is outside the range of data collected in its training or open-loop mode. The hybrid system provides a linear system that takes over control when parameters are outside the neural network's range.

A power plant near Warsaw, Poland, recently installed and tested such a hybrid system. The Ostroleka power plant near Warsaw, Poland, is one of the first to apply a hybrid neural network MPC system with an Aspen Target MPC controller using a nonlinear state-space model integrated with a neural network model.

The plant uses three wall-fired boilers with pulverized coal. Six sets of low NO_x burners are installed in each of four elevations. After a period of open-loop training the MPC was placed in closed-loop operation with the following results:

- NO_x decreased 15 to 25%,
- Boiler efficiency increased 0.1 to 0.5%,
- LOI (carbon loss on ignition) decreased from 5% to 3%, and
- Cooling water consumption reduced by up to 5%.

The controls also demonstrated robust response to extraordinary events.

Burner Management Systems

To ensure that modern power plants have the capability to respond promptly to any load demand, new burner management systems (BMSs) provide the programmable control needed to operate burners and igniters for gas or oil firing. They also provide the highest available level of safety for automatic control and monitoring of start-up, operation, and shutdown of the boiler.

In Melbourne, Australia, a new BMS was installed at the Newport Power Station. The strategy was to upgrade the 1980s stand-alone combustion controls using a new distributed control system (DCS). The tight integration planned for the two systems was simplified by the flexibility in the BMS design, which also can operate independently of the DCS.

There is a total of 60 burner elements, with each element requiring individual control and monitoring with a high level of requirements related to safety. A QUADLOG safety system from Moore Process Automation Solutions was used. It has a one-out-of-two-diagnostic (1oo2D) redundant architecture. In the 1oo2D redundant architecture, there are two processors, each with its own backplane and infrastructure.

The PC-based programming software runs on Windows NT and provides an intuitive, user-friendly graphical configuration. There is a choice of several programming languages: ladder logic, function blocks, sequential charts, or structured text. The graphical configuration capability saves hours when making on-line changes. The system is fully self-documenting and the high availability of the system results in higher plant availability.

If the on-line diagnostics detect a failure in one system, the other system automatically assumes control and the safety system remains available. The control and I/O modules fit into slots with 10 to a rack. There are four racks assigned to purge and fuel safety, and 8 assigned to the burners and igniters. The racks are connected to a LAN, along with a workstation located near the racks and 2 workstations located in the control room.

Each of the 20 gas burners, 20 oil burners, and 20 igniters has its own flame detection system, valves, limit switches, air dampers, and other equipment. The overall safety system has four burner subsystems:

- igniter fuel trip,
- gas fuel trip,
- oil fuel trip, and
- master fuel trip.

All failure modes are predetermined in the hardware and firmware to a fail-safe condition. Alarms are used to identify them.

The application software for sequencing operations allows a single button to initiate the following automatic actions:

- purge the furnace,
- perform comprehensive header leak tests on all the fuels,
- open the trip valves,
- bring the first igniter and burner on-line, and
- bring the furnace up with additional burners as needed.

A saturated steam curve ignition run-up is used and burners are bypassed that fail to go into service. If a boiler is cold, it is pre-warmed, firing it at a level according to the cold-boiler-run-up steam curve. These actions speed the process of responding to load demands. The burner management system has algorithms that assist in the required configurations.

References

Fuller, James L., *Robotics: Introduction, Programming, and Projects*, Merrill, an Imprint of Macmillan Publishing Company: New York, 1991.

Harvey, Charles and Peter Wise, "New BMS and DCS Provide Reliable and Responsive Power Plant Automation," *Instrumentation & Control Systems*, Vol. 72 No. 11, November 1999, pp. 59-61.

Redman, Jim, "Upgrades Enhance Capability," *Industrial Computing*, Vol. 18 No. 5, April 1999, pp. 22-25.

Rinehart, R.R., "Model-based Control," *Instrument Engineers' Handbook*, 3rd Edition, Chilton Book Company: Radnor, PA, 1995.

Riorda, Michael, "The Incredible Shrinking Transistor," *Technology Review*, Vol. 100 No. 8, November/December 1997, pp. 48-50.

Zink, John C., Ph.D., P.E., "Power Plants Merge Information and Control Systems," *Instrumentation & Control Systems*, Vol. 72 No. 11, November 1999, pp. 55-57.

Chapter 7

Computer Networks And Security

During the early days of the computers in the mid-1960s, the computer environment consisted of large mainframes with remote terminals connected to the mainframe. These large computers required special environmentally controlled computer rooms. Both the computers and the rooms were expensive to build and required trained personnel for input and output and maintenance. Later, these large computers would grow smaller and some applications would be served by minicomputers that did not need special rooms. The remote terminal system was the main user-computer interface until around 1980, when the first personal computers were introduced.

The early methods of computing had many disadvantages as well as some advantages. The computing environment allowed multiple users to access one computer. These users had access to the same applications and files and had some capability of transmitting and receiving mail electronically. Centralized management of the computer was possible. The management of user accounts, files and the hardware was done by a group of specialists. The costs of running the cables for the user terminals were the second most expensive part of the computer installation besides the cost of the computer itself.

This computing environment was known as centralized computing. If the mainframe or minicomputer went down, all of the users were affected and would remain idle until the computer became operational. These breakdowns were common especially where the terminals were separated at some distance from the computer. The breakdowns were costly since the users were unproductive until the computer was operational.

The user terminals were known as dumb terminals, since they had very little intelligence with most of the host-terminal interface processing done on the mainframe.

Multiple-host connections were possible using a modem to dial up another computer. This centralized computing environment was used because at the time other alternatives were not available. In 1981 IBM introduced the personal computer and the computing environment was changed forever.

The personal computer gave the user much more freedom. The personal computer brought the mainframe functionality to the desktop. Initially, it did not bring the power of the mainframe to the desktop since the early PCs had very little power.

PCs became popular because they did away with the disadvantages of a centralized computer including the cabling costs, environmentally controlled rooms and expensive hardware and maintenance costs.

The user had control over their computing environment and enjoyed having personal application software that would run on individual computers. A user could load an application program, enter the data and print out a report on a local printer much quicker than before. There were no multiple-user applications, messaging capabilities or multiple-host support.

Mainframe-PC connectivity was accomplished using terminal emulation programs that were run on the PC to emulate the terminals. File transfers between the mainframe and the PC were supported with the terminal emulation software.

File transfers were also done using a technique known as Sneakernet. Here, a user would save their work on a floppy disk and take or send the disk to another computer and load it.

Shortly after the introduction of the personal computer, the first local area networks (LANs) started to appear. These early LANs allowed a connection in the host-terminal environment and also interconnection with the personal computers. Devices and peripherals could be shared on a single cabling scheme.

At first, most LANs were relatively expensive to install and maintain. They started in scientific and engineering applications and spread to almost all commercial activities. Maintenance costs were also higher with the early LANs since few people knew how to install and maintain them.

The disadvantages of LANs were gradually reduced. Cable costs were reduced and in small and medium-size networks, these cables could be installed by the user. Networks usually transmit and receive in megabits per second (Mbps) which is much faster than the 9600 or 19,200 bits per second (bps) that the terminal-host interface supported.

International Standards Organization Model

The standards association known as the International Standards Organization has developed an architectural model known as the ISO reference model for open-systems interconnection. This model divides a local area network system into seven processing layers. Each layer performs specific functions as part of the overall task. The application programs can run on different computer systems which can be located anywhere. These programs can communicate with each other as if the programs resided in the same system.

The ISO model is based on modules and is not specific to software or hardware. ISO defines seven modules but does not cover the software code or hardware that implements the model.

Local area networking protocols such as the Transport Control Protocol/Internet Protocol (TCP/IP) and hardware protocols are designed using this model. These protocols may be intermixed completely. The ISP model design allows this.

There are seven modules or layers in the model; physical, datalink, network, transport, session, presentation and application layers. These seven layers are shown in Table 7-1.

Table 7-1. ISO Reference Model—Open-system Layers

Layer	*Function*
7 Application	User applications, process management
6 Presentation	Data interpretation, format, code transformation
5 Session	Administration, control of sessions between nodes
4 Transport	Transparent data transfer and transmission control
3 Network	Routing, switching, flow control
2 Data link	Maintain, release data, link, error, flow control
1 Physical	Control of data circuits, physical media definition

Each layer has a specific purpose and functions independently of the other layers. Each layer is aware only of its immediate upper and lower module.

The modularity of the ISO model design means that each individual module can be replaced with a module of a different protocol. The new module must provide the same functionality as the module it replaced. This is accomplished without any interruption in the function of the other layers. This approach allows multiple-vendor operation and encourages technology changes.

Functional Layers

The functions performed at each layer of the ISO model define the different types of hardware and software needed to install or enhance the network. The physical layer defines the lowest or first level. This layer defines the methods used to transmit and receive data on the network. It consists of the wiring and devices that are used to connect a station's network interface controller to the wiring and the signals needed to transmit and receive data on the network.

The data-link layer is the second layer. This layer synchronizes the transmission and handles the packet error control and recovery. Frame formatting and redundancy checking to check for errors in the data are accomplished at this layer. This layer defines the network access method such as Ethernet.

The network layer controls the forwarding of messages between stations. This layer allows the data to flow sequentially between stations in the most economical path both logically and physically. This layer also allows packets to be transmitted over devices known as routers.

The transport layer provides end-to-end or station-to-station transmission of the data. This layer is responsible for the transmission of data between communicating stations.

The session layer establishes, maintains and disconnects the communications link between two stations. This layer is also responsible for the station physical addresses.

The presentation layer is responsible for the data format and data encryption. It is not always implemented.

The application layer is the seventh layer. This layer is used for applications that run over the network. These applications include file transfers.

Since each layer has its particular role, it provides structure in what is a static model. Each layer has its place and cannot be moved to replace another layer above or below it.

Networking Technology

Networking technology includes the simplest local area networks to those that span large, multiple-site organizations. An organization needs to leverage the creative uses of information and networking technology for its strategic benefit. There is a continuum of growth with computer and communications technology. The networking trend is continuing. Each success causes more complex and intuitive applications of networking to sustain an advantage.

The proliferation of communications technology has redefined the path of success for organizations. Communications have become integral to most computer and business activities. Some will fail to recognize the available opportunities, and others will fail to recognize how to adapt them to the new possibilities that will be available.

Technological changes will greatly improve the efficiency and enhance the implementation of control and communication systems. Networking, in particular, can be a sophisticated strategy that can lower operating costs and improve work environments.

Networks can facilitate quick decisions, improve information flow and accuracy, and communicate such information and decisions rapidly to those who would benefit from them.

Networking provides an economy-of-scale since it distributes operations to smaller, less expensive, and more responsive personal computers that are interconnected.

Networking also provides shared access to data for many users that were previously unavailable. The technology shifts from workstations to more common personal computers.

Information and networking technology are strategic solutions, but this technology must be applied with foresight, forethought and a critical understanding of the technology. Clearly, there is a risk to try new technology with its potential for clumsy and inefficient results. But, there is a greater risk for those who watch and wait while others succeed first.

Advantages of PC Networks

PC networks yield significant advantages. Low cost networking power can provide more efficiency with fewer resources. The technology of distributed personal computers is very cost-effective for many functions.

Networking can streamline the control of systems that are inherently slow or fragile and automate these systems for higher integrity. It is proven that networks do generate cost-effective benefits, create significant economies of scale, boost productivity and create unanticipated and imaginative results. Such systems have generated information that has been applied to pare costs, simplify excessive or expensive procedures, locate cost variances and reduce inventories and stocks.

The control network must be segmented and protected from business traffic to ensure the required reliability and determinism. Simply separating control and business networks with a bridge, or even an NT server, is rarely sufficient. An incorrectly configured Windows workstation can generate high levels of broadcast packets that can affect the sample time. Security breaches through the control network can also cause problems.

DDE/OLE Linking

Dynamic Data Exchange (DDE) was designed as an MS Windows tool for linking data within a single PC, but many software vendors have created extensions for full network DDE data access and linkage. This includes network data and data accessible through phone connections, remote sites and enterprise-wide networks.

Object Linking and Embedding (OLE) provides a means to link data cells and the latest versions of entire documents, images, video and sounds. The availability and prominence of interconnectivity, interoperability and the development of user-level tools increases the importance of existing networks.

Ethernet Trends

The future of Ethernet is strong and it will not merely be bypassed by technological changes. Other network schemes like Carrier Sense Multiple Access with Collision Detection (CSMA/CD) networks can provide better response times, throughputs and lower overhead, but this is offset by the consistency and predictability of token protocols such as Ethernet.

Ethernet was originally thought to reach its peak in about 1990 and become obsolescent by 2003. However, this is one network technology that has blossomed and continues to gather momentum past its planned half-life.

Ethernet is doing well in the industrial marketplace. It has completely taken over the commercial sector, pushing the former competition like ARCnet and token ring aside. Users and vendors have given Ethernet a place in modern control systems.

There will soon be an estimated 200 million Ethernet connections worldwide. Some control networks may not survive the growth of Ethernet, including controller-to-controller buses such as ControlNet or Profibus-FMS, unless they are modified to use Ethernet.

It has become a very inexpensive technology and is well understood since it is one of the simplest network technologies to implement (Figure 7-1).

Ethernet offers the possibility of a truly open connectivity standard for distributed control systems along with field-device networks such as Profibus or Foundation Fieldbus.

Ethernet technology was originally developed by Xerox in the early 1980s. It was adopted by the Institute of Electrical and Electronics Engineers (IEEE) as standard 802.3 in 1988. Since then it has become the dominant standard for local area networks.

Ethernet provides the physical layer and the data-link layer protocols in the ISO model. The physical layer defines the cable types, connec-

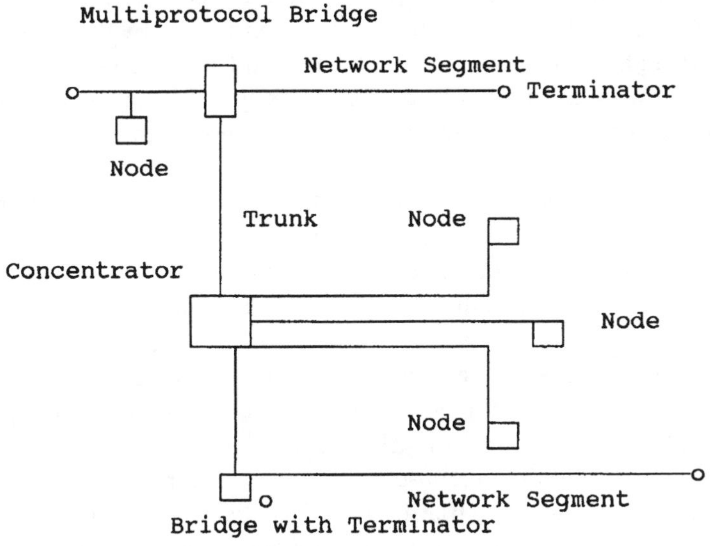

Figure 7-1. Basic Ethernet network

tors, and electrical characteristics. The data-link layer defines the format for an Ethernet frame, the error-checking method and the physical-addressing method.

The data-link layer also defines the protocol Ethernet uses to determine when nodes can transmit on the network. This is known as carrier sense, multiple access with collision detection. There is a need for standard upper-level protocols to be used with Ethernet.

Ethernet is only one of the protocols needed to make any LAN operational. Ethernet cannot help a message find its way through a complex network. It cannot define how to carry out specific tasks on a network, such as file transfers. These operations require additional protocols on top of Ethernet to create a protocol suite.

In Ethernet there is no assurance of when a node can get on the network. If the network is busy, then a node with a message may have to wait. The length of the wait would be based on probabilities. This makes Ethernet nondeterministic.

In a proportional-integral-derivative (PID) control system, it could be a problem. The collection of sensor data over an Ethernet segment should have a regular sample time to be effective. In theory, Ethernet may not be able to guarantee consistent timing.

The determinism issue would have you believe that any Ethernet network loaded at about 40% would encounter an exponential growth in transmission delay times. But, studies conducted in the late 1980s showed that, in practice, Ethernet delays tend to be linear and can be held to less than 2 milliseconds for a lightly loaded network and 30 milliseconds for a heavily loaded network. These delays are acceptable for most HVAC control applications.

The main concern is to keep the traffic level on the Ethernet network low enough so that collisions occur at low frequency levels. This means restricting the number of devices and the amount of traffic that each device generates so that the total segment traffic never rises above 1 or 2%.

If more devices are needed, they are put on separate segments and then connected using a bridge, router, or switch. This concept is used by DCS systems such as ABB's MasterBus and Foxboro's Nodebus.

Besides the commercial-grade hardware for Ethernet, several manufacturers offer industrial hubs and switches that can be DIN-rail mounted and have redundant power supplies. Ethernet is easy to install and connect.

Ethernet Protocols

Since Ethernet is not sufficient in itself to run a network other protocols need to be used. The transmission control protocol (TCP) and Internet protocol (IP) are the protocols of choice for dealing with issues of routing and end-to-end data integrity. These are the network and transport layers in the Open Systems Interconnect (OSI) protocol model.

While TCP and IP are transport and network layer protocols, they usually come bundled with application layer protocols like file transfer protocol, simple mail transport protocol, and simple network management protocol (SNMP). Several control systems vendors use some of these protocols (such as SNMP) in their hardware, but they are not ideal for control systems. The MODBUS application protocol can be used. This is the application layer from the MODBUS protocol suite, bundled with TCP/IP.

Both TCP and MODBUS are widely supported. However, MODBUS does not support an object-based communications model like most of the new fieldbus technologies. When you read a MODBUS device over a network, you only get the data without any details about its function or format.

It is no longer enough to consider a device as a group of registers. The network must be able to see an object. The IEEE 1451 standard provides techniques for self-describing measurement and control devices. It is an object technology that specifies a common object model for devices, how they communicate, either client/server or publish/subscribe and how they are managed.

These network-neutral interfaces can be implemented on Ethernet and they provide a way of integrating multiple networks and using common application software.

The Fieldbus Foundation uses fast Ethernet as the base-level protocol for its H2 network. The foundation maps the Layer 2 technologies from its H1 standard into Ethernet. This includes the object model and scheduling and publisher/subscriber services. Other systems encapsulate their protocols inside of an Ethernet frame. This technique is known as tunneling.

The utility of Ethernet has grown with the wide acceptance of the Transaction Control Protocol and Internet Protocol (TCP/IP) suite. Part of this success has been due to the standardization and the multiple platform support provided by Ethernet. This success is in contrast to the lack of vendor conformity in other network protocols such as StarLan, ARCnet, and Token Ring. Some of this success can also be attributed to the growth and economics of twisted-pair and optical fiber technology which have extended the life of Ethernet.

The high costs of going with the wrong emerging network technology also encourages users to be more conservative and choose the path of known and accepted technologies like Ethernet. This conservatism is another major reason for the success of Ethernet and its continued applicability.

Ethernet economics also continues to improve. Ethernet hardware which originally cost thousands of dollars is now available at less than 10% of that cost.

Ethernet is a description of physical media interconnections. Ethernet is linked to TCP/IP and some users consider Ethernet to be TCP/IP, but both exist independently. It is likely that Ethernet processes will be separated from the underlying medium much as TCP/IP has become a protocol independent from the physical transmission process.

Ethernet and TCP/IP provide a strong foundation for linkage. They provide an efficient method to connect dissimilar computers, operating systems and network protocols.

Existing Ethernet networks are not likely to be replaced by faster or merely new technology. Instead, they will be integrated into wider networks. This type of data access and linkage has been pushed by the utility of PC and Windows-based products. These tools provide uniform systems for users to gather, construct and process networked data.

Software vendors now build applications that are virtually independent from the underlying network protocols including Ethernet. The use of compound networks, backbones, interconnectivity, bridging and routing will increase.

Ethernet remains in a growth mode because the tools and experience exist. It will not be replaced for years, but will be supplemented when specific situations demand higher transmission speeds. The installed base and compatibility ensure its survival for many more years. Ethernet can also be repackaged to delay its obsolescence.

Ethernet persists because of its simplicity. The twisted-pair star configuration utilizes telephone wiring to simplify installation and lower costs.

The Open Systems Interconnection (OSI) communication structure provides the means to alter layers independently of the others. The OSI structure of most high-level protocols allows a rethinking, rebuilding and rebundling of network services. This repackaged Ethernet technology also provides a starting point for small network installations that can gradually expand to a coaxial backbone connecting concentrations of twisted-pair installations.

Ethernet provides plug-and-play intercompatibility that is otherwise unavailable. This may be sufficient to prevent the replacement of existing networks.

Expect to see more implementations that encapsulate Ethernet. Implementations of routers, packet switches, and gateways can boost the integration of small networks into wider enterprise type systems.

Ethernet comprises layers 1 and 2 of the OSI model. Layers 3 through 7 are not part of Ethernet, although Ethernet is often packaged with the TCP/IP protocol. TCP/IP provides the network-layer functions.

TCP/IP was commissioned by the Department of Defense for the Internet wide area network (WAN) network. It was standardized by the government, but is supplied in a rapidly growing list of products. The advantage of TCP/IP is its simple structure and its ready implementation within the memory and speed limits of most computer platforms. The disadvantage of TCP/IP as implemented is that it does not conform rigorously to the OSI protocol.

Ethernet is represented by hardware but software protocols build upon the basic hardware that defines Ethernet. Ethernet is physical, although most Ethernet networks include network software such as TCP/IP. This difference between the OSI model and Ethernet TCP/IP is apt to fade.

The physical layer acts as a conduit or data path and the data-link layer is implemented by software, controller and transceiver units. The software function is often serviced by TCP/IP and is represented in the data-link layer by communication control protocols.

Ethernet Variations

Ethernet hardware has several variations and there are physical differences among them (See Table 7-2). There are standards for the various transmission media, including baseband, twisted-pair, radio frequency and infrared, broadband coaxial cable and optical fiber.

Baseband coax cable, single-strand and single-channel has been the prevalent medium.

Table 7-2. Comparison of Ethernet Networks

Twisted-pair Network
- Less expensive and easier to install than coaxial cable
- Simple node relocation
- Smaller geographic area
- Less tolerance to crosstalk, interference
- Fault isolation more difficult

Coaxial Cable Network
- More tolerance to crosstalk, interference
- More durable and reliable than twisted-pair
- Harder to install and more expensive than twisted-pair

Network with Trunk Backbone
- Better performance than cascaded network
- Simple expansion
- Can segment failed networks
- Easier to troubleshoot
- Harder to trace

Wireless Network
- Simple to install
- Node relocation simple
- Can be moved from site to site
- Limited distance
- Slower transmission speeds
- Possible health consequences
- Susceptible to eavesdropping
- Poor tolerance to interference

Coaxial cable has a conducting outer metal shield enclosing and insulated from a central metal conducting core. It is called coaxial since the core and shield conductors share a common axis. A single-strand cable with a tinned copper core is used which is surrounded by a foam material that insulates it from a tinned and braided copper shield. Coaxial cable provides an electrically balanced signal. A single coaxial cable transmits a single channel of information.

A broadband coax cable is similar in function and appearance to baseband, although the cable is physically several times larger. Broadband Ethernet, in coax and optical fiber, will provide longer network segments and increased channel capacity or channel transmission speed. It is called broadband since it provides a wider frequency range than baseband. There is multiple-channel broadband Ethernet cable, both copper and optical fiber. Various data transmission and encoding schemes are used in Ethernet. The broadband signal characteristics are shown in Table 7-3.

Table 7-3. Broadband Signaling Characteristics

Broadband Signaling Characteristics
Many channels on one cable
Separate frequency for each channel
Carrier tone supplied by modems
Modulation encoding schemes
AM amplitude modulation
FM frequency modulation
PM phase modulation
PCM pulse-code modulation
PSK phase-shift keying
FSK frequency-shift keying

Baseband cable includes 50-ohm Thicknet coaxial cable, 75-ohm Thinnet and Cheapernet cable. The higher cable impedance of Thinnet and Cheapernet limits them to shorter segments because the signal degrades and attenuates with distances. The thinner cable is also more susceptible to signal interference.

Thinnet is a simple cable substitute for standard Ethernet network cabling, while Cheapernet uses a different Ethernet controller-transceiver unit that is integrated in a single bus card or included on the PC mother board.

Optical fiber cables are thinner, lighter and contain no metal. They are used to solve long-haul or security problems. Fiber can be used to interconnect buildings miles apart.

Broadband cables are most often used for analog communications, but optical fiber broadband is used to transmit digital transmissions as well.

Broadband networks provide more bandwidth, higher speeds and a larger geographic service than baseband networks, but the cost is higher. The physical medium and transmission differ from baseband but the Ethernet protocols can transmit on baseband and interconnect using a bridge.

Broadband can simultaneously carry voice, data, video or any other digital or analog traffic at different frequencies. When broadband networks carry voice, video, and data communication signals concurrently on the same networks, a bridging mechanism is required to transmit data between networks using different transmission media. Broadband disadvantages include its initial expense as well as the difficulty and expense in moving or adding nodes to expand the network.

The main difference between a standard Ethernet, twisted-pair, and the Cheapernet and Thinnet variations is the maximum length of cable each version supports and the minimum distance between nodes.

Standard Ethernet supports a maximum 500-meter segment and nodes spaced no closer than 2.5 meters, while Cheapernet and Thinnet run on thinner 75-ohm cabling and support a maximum 200 meter segment. The nodes can be spaced as close as 0.5 meter. The 50-ohm coaxial cable is physically thicker and stronger, better shielded, and more expensive than the 75-ohm Thinnet cable. Broadband uses a television tuner-like modem that converts each channel to a separate frequency.

There can be 2 to 80 channels per broadband cable. Broadband Ethernet can run for longer distances than baseband, up to 2 km. Optical fiber technology is newer and provides greater bandwidth and higher transmission speeds than broadband coaxial cable, but it is more expensive.

Fiber cable is more difficult to splice and repair. Expansion and contraction from extremes of heat and cold can fracture the cable. Heat and cold can also affect the optical qualities and gradually decrease the bandwidth.

Optical fiber transmission is available through PC adapter boards for Ethernet. Optical fiber networks have some of the same transmission characteristics as broadband cable but they are more expensive and more difficult to debug, maintain, and expand. They are not susceptible to electronic noise or signal crosstalk and provide good security.

Thinnet and Cheapernet cables have similar compositions. The dielectric is a type of foam and the exterior insulating jacket can be polyvinyl chloride (PVC) or the more expensive, fire-resistant fluorinated ethylene propylene (FEP) Teflon for installation in building plenums and walls.

Baseband Ethernet can run on twisted-pair telephone wiring. This type of cable has bundles of wires containing 4, 6, or 8 wiring pairs which are twisted around each other. The wires are twisted to minimize crosstalk, signal attenuation and susceptibility to stray noise (See Tables 7-4, 7-5).

Table 7-4. Ethernet Baseband Variants

Specifications	*Ethernet*	*Cheapernet*	*Thinnet*	*Twisted-Pair*
Segment length (m)	500	200	200	250
Transceivers per segment	100	100	100	512
Maximum number of segments	3	3	3	3
Network length (m)	1500	600	600	500

Table 7-5. Baseband Signaling Characteristics

One channel per coaxial cable
No carrier
DC to upper frequency limit
Encoding schemes
- Manchester
- Return-to-zero
- Non-return-to-zero

Digital transmission, no modem translation

Ethernet Compatibility

Network compatibility is a function of network length and length of individual node connections. Ethernet transmission is based upon a statistical signal and ranges of signal acceptance. Violation of those specifications means that the network could function poorly. Exceeded segment lengths, drop cable lengths, twisted-pair connections, substituting Thinnet for standard cable segments or inserting twisted-pairs all stress the network.

Some Ethernet versions have data encoding differences. The OSI model has become generally accepted due to the globalization of telecommunications and data communications. It is a de facto standard. True standardization implies comprehensive testing for conformity to OSI and this is a major undertaking in the future.

Wireless Ethernet

Baseband wireless networks use infrared or radio frequency signal transmission to transmit signals through the air and through the walls. Spread-spectrum technology (SST) uses multiple frequencies and carries a single baseband channel. The only difference between the more traditional cable networks and a wireless network is the lack of media. No connecting wires are used.

This technology is useful for buildings where adding wiring would be costly. This includes old buildings without conduits, hollow walls, or false ceilings, or buildings with an asbestos problem.

Infrared technology is not advisable for dusty, dirty, or smoky environments because these particles degrade the signals. Infrared technology is also limited by line-of-sight installations.

The hardware required for the most minimal configuration is node-based wireless transmitters which eliminates the need for any network cabling. The more usual configuration uses a backbone cable or wire that interconnects the nodes. The cable or wire is either coaxial cable, twisted-pair wire bundles and telephone modular jumpers, or optical fiber.

Coaxial cable is most often designated as RG-50, RG-59, or RG-225 coaxial cable. The 50-ohm cable (RG-50) performs better than 75-ohm CATV cable (RG-59 or RG-225) because it has better resistance to low-frequency electromagnetic noise which means that it shields the signal and is rarely affected by external interference. Thinnet and Cheapernet are 75-ohm cable and can suffer this interference.

Unshielded twisted-pair (UTP) Ethernet wire is designated as either UTP-3, UTP-4, or STP-3, which is a higher-grade shielded twisted-pair wire bundle. The wire itself is typically 24-gauge copper wire with a PVC or Teflon coating.

An Ethernet network consists of the cable and connectors. There are also terminating resistors, coaxial taps and transceiver drop cables. Transceivers and controllers are used at the node locations. Fan-outs, gateways, repeaters and other equipment are used to expand the system. An Ethernet gateway provides network expansion and conversion among networks with different protocols.

A packet or Ethernet frame is the transmission vehicle. Ethernet acts like a one-line country telephone network since only a single transmission can proceed at a time (Figure 7-2). The transmission protocols are rules for deciding who can transmit.

Each node has transmission and reception hardware, which takes the role of a telephone handset. This hardware controls access to the communications channel. Each node builds its message to match the required Ethernet frame format.

Transceivers and Controllers

The transmission-reception hardware is called a transceiver, or transmitter-receiver. This is analogous to the earphone and microphone of the telephone headset.

The traffic control hardware is called an Ethernet controller, which is analogous to the dial unit of a phone system. The controller and trans-

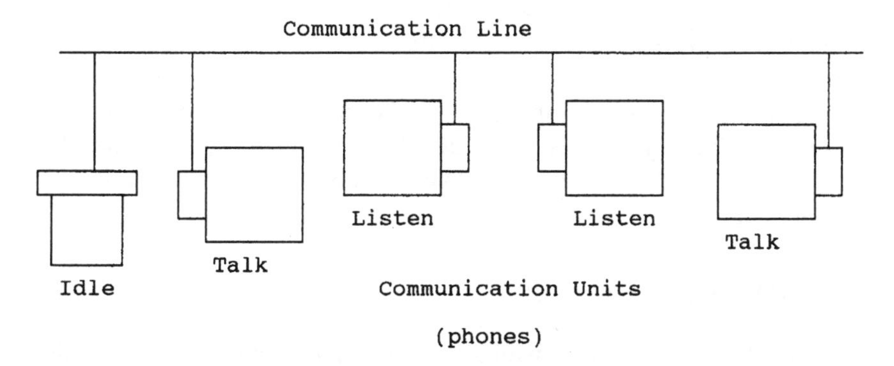

Figure 7-2. Party-line telephone analogy

ceiver are often bundled together into a single personal computer adapter board called a network interface card (NIC) or a multimedia access unit. Transceiver and controller functions are specified by the OSI model. Table 7-6 lists the transceiver functions.

Table 7-6. Transceiver Functions

Table 7-6. Transceiver Functions
Receive signal
Transmit signal
Broadcast jam signal
Test for reception
Test for transmission
Test for jam (collision)
Test for heartbeat
Test for carrier sense
Transmit preamble
Transmit delay signal
Sense collision

The transceivers limit access to one caller at a time and direct the transmission. First, the transceiver listens for a quiet period. When the network cable is not busy with another transmission, the transceiver transmits the data signal.

Two or more transmitters may simultaneously observe a quiet period and transmit. This results in a collision. Then, the transmitting transceivers will broadcast a jam signal. Upon receipt of this signal, each station contributing to the collision must wait for a random period of time before attempting to transmit again. Different mechanisms are used for collision detection. These differences make some hardware incompatible with other hardware.

The Ethernet controller dials the number and builds the transmitted packet which is addressed to a designated node. The controller also disassembles packets and transfers this information to the receiving computer.

The network access unit (NAU) also provides some services to increase reliability. The NAU shuts down when it senses that the twisted-pair wire is connected to a live telephone circuit. It also tests the connections.

Transmission Process

Baseband Ethernet transmits a single digital channel. A base voltage of + 0.7 V provides a carrier sense signal for the nodes on the network. When transmitting, a transceiver unit lowers the voltage level to -0.7 V. The transmitting node propagates a digital signal wave outward in both directions down the coaxial cable.

Manchester encoding is used where the digital signal represents a 0 as a high voltage with a transition to a low voltage and a 1 as a low voltage with a transition to a high voltage.

Twisted-pair concentrators interpret and regenerate the signals for retransmission to other nodes. This integrity check is not available on coaxial networks. A short circuit or other electrical malfunction will cause the failed node to be removed.

Packet Transmission

Individual bits are not transferred from location to location without a preamble or explanation as they are in modem-connected lines. The information to be transferred is packaged and checked upon receipt for proper delivery.

The packet contains the following information:

- destination address,
- return or source address,
- length field,
- data field and
- frame check sequence field.

Handshaking between transmitting and receiving nodes prevents packets from crashing into each other and being scrambled. When the network is not busy, an idle signal is transmitted by all source-destination nodes, which is the -0.7 V carrier sense. This base voltage is sometimes called the heartbeat.

Once the data set is packaged into a frame, the Ethernet controller polls the transceiver for the network state. If the network is busy, it waits until the network is free and then transmits the packet. The packet signal propagates in both directions over the coaxial cable.

Each transceiver on the network hears the transmission preamble, synchronizes its clock, and awaits the packet. This clock synchronization is needed to distinguish the beginning and the end of the packet. The preamble provides this frame alignment.

Every transceiver on the network reads the address information. When a packet destination address matches the Ethernet address of a node, the transceiver for that destination node accepts the Ethernet signal into the receive buffers on its Ethernet controller. The signal wave continues traveling down the cable until absorbed by the terminators.

All transceiver units listen to the coax broadcast signal. When the destination address matches the Ethernet address of a receiving node, that transceiver reads the signal and copies the transmission into its own buffer without changing or damaging the signal wave.

The controller verifies that the length of the data message matches the value in the length field. Any discrepancies are passed to the software protocols. TCP/IP, for example, would request a rebroadcast of the damaged packet. TCP/IP also reports receipt confirmation to the transmitting node. If the transmission has been successful, the data field contents are passed to the software protocols and eventually transferred to the application.

Ethernet Drawbacks

The nodes contend for network access, much like users of party-line telephones compete for a free line. This contention method of broadcasting allows more than one transceiver to transmit at the same time. Two or more transceivers may simultaneously sense a quiet network. In that time interval, each will begin to transmit packets. The signals will collide, corrupting each transmission. One unit may recognize the collision and broadcast a jam signal. As a consequence two fragmentary packets are transmitted in each direction for each corrupted transmission.

When a collision has been sensed and acknowledged, the Ethernet controllers create a random number. This is usually provided by the node computer equipment and is used to delay further retransmission attempts. The random delay serves to restart the contention for the network.

These collisions on Ethernet create fragmentary packets and lower the possible transmission rate. The jam signal takes a minimum of 32 bits and all colliding packets must be retransmitted.

An administrative layer directs the packets and deals with traffic problems from packet content. This software provides the ability to know what nodes exist. Some systems map network station names into logical Internet addresses. The role of this administrative layer is similar to that of the telephone operator. The operator provides connections, distributes telephone numbers and verifies wrong numbers, retries the connection after a busy signal, switches connections onto other networks when the main network fails, troubleshoots the mechanical connection and informs the calling party of network problems.

While hardware problems are more easily diagnosed, network software problems can be narrowed to specific nodes with software tools and protocol analyzers.

Network software can provide functions such as system boot, network initialization and systems monitoring. Simple networks build upon TCP/IP protocols which are the predominant Ethernet software. These provide intercommunication between nodes for messages, file exchange and terminal services. This can include remote workstation operation, remote boot, and remote device access and services.

Internet Protocol

The Internet protocol had its roots in a communications addressing scheme known as the Address Resolution Protocol (ARP). This formed the basis of the Department of Defense's Advanced Research Project (DARPA) four-node network called Arpanet.

Arpanet evolved into the Internet during the 1970s as TCP/IP was adapted for LANs using Ethernet. The IP address contains a network and host address divided into parts.

The Internet protocol provides the lowest level of software access and interface to the Ethernet environment. This protocol provides link initialization and termination, node recognition, and packet control operations.

Transmission Control Protocol is a higher level communication protocol. This protocol sequences data transfers and the actual packet transmissions. TCP resides at the OSI layer 4, while IP is OSI layer 3.

Internet-enabled Networks

This type of network makes it possible to manage, view, and control processes, network devices, and configurations from applications and standard Web browsers. Implementing a Web network allows visualization and control over the Internet via technologies such as IIS/Peer Web server, HTTP, XML, ActiveZ and embedded Web servers (Figure 7-3). This is sometimes called a Web-Aware Enterprise Network (WAEN).

The recent explosion of WAENs in industrial automation has caused some concern about the security of networks connected to the Internet. Implementing a security blueprint that fits your Internet requirements is critical as Web capability moves ahead. Several steps can be taken to secure a network against unwanted intrusion when connected to the Internet. There are some basic security technologies that can be used against the commonly used attacks that could compromise a network's integrity.

Connecting a LAN or WAN (Wide Area Network) to the Internet exposes the network to more than 50,000 other networks and their users. These connections provide opportunities for information sharing, but

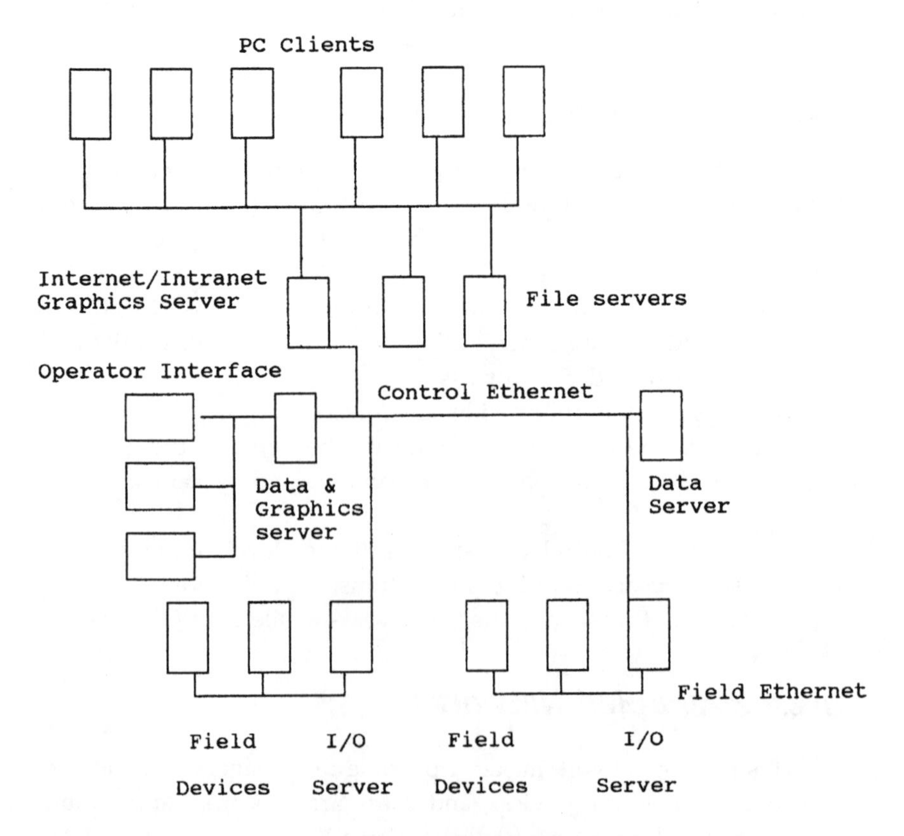

Figure 7-3. Browser-based system

they also lead to information theft or corruption. Network information can reside on a hard drive or be in transit across the network cable in the form of packets. Serious hackers are also called crackers. They may become active in accessing your network and should be considered a threat. Some of their methods create holes within the operating system, which allows easy access for hackers.

Security Technology

Levels of security should be implemented for Internet implementations. These will restrict the levels of access made available to remote users. This type of security is available from NT, Internet Information Server (IIS) security, SSL (Secure Socket Layers), Digital Certificates and encryption.

In Wonderware software there are three types of protected data access via the Internet. These are:

- Read-only access of SQL clients, which are ActiveX objects,
- Read-only access of real-time process graphics using standard NT and IIS security,
- Runtime, a module which gives remote users the ability to respond to emergencies, security for the user is predefined using NT and IIS security administration.

In Windows NT-based systems, general security is supported through the OS and IIS (Figure 7-4). Windows NT allows the user to enable and disable access to files and directories by user or group. With IIS manager more security restricting read/write accesses can be added.

Security technology includes SSL, Digital certificates and HTTPS. SSL is a protocol developed by Netscape that allows a secure transaction between standard browsers and Web servers on the Internet. Both the Netscape Navigator and Internet Explorer browsers support SSL.

Encryption is another type of security. Data can be encrypted using a secret key for encryption (symmetric) or public key encryption (asymmetric). In secret key encryption both parties share the same password or key. In public key encryption, there are two keys that are related, one is private and the other public. SSL uses the public key method of encryption.

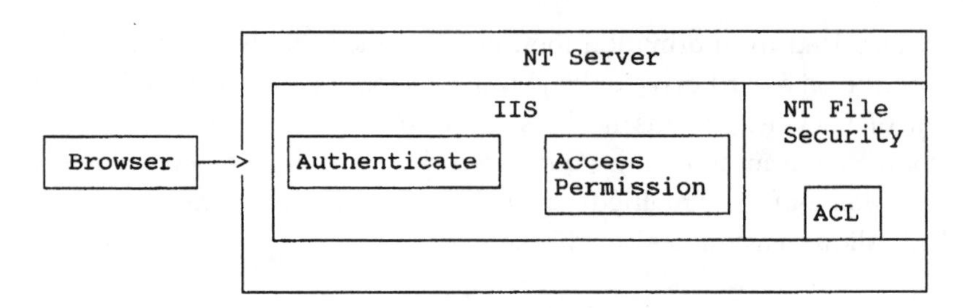

Figure 7-4. Security supported by Windows NT and Microsoft IIS Web server

Digital Certificates

Public keys can be authenticated with Digital Certificates (IDs). When a Digital Certificate is used with encryption, it secures the authenticity of the parties involved in the transaction (Figure 7-5).

Digital Certificates provide a type of identification or passport that allows users to navigate across multiple networks. The users retain their access levels and credentials (Digital IDs) for each network. Digital Certificates provide the capability to identify users, access levels and authorized functions in multiple networks. When a user's levels are defined and a digital Certificate is issued, administrators can audit the user's transactions.

Methods of outside intrusion that can compromise information and security on a network include:

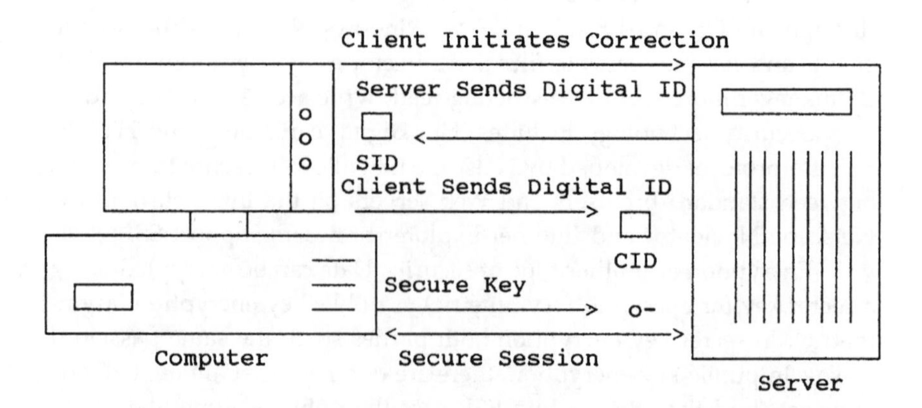

Figure 7-5. Secure transmissions with digital ID certificates.

- network packet sniffers,
- password attacks,
- IP spoofing,
- middle man attacks,
- denial-of-service attacks, and
- application layer attacks.

Packet Sniffers

Networked computers communicate serially by sending one data piece after another. The large data pieces are broken into smaller pieces. The smaller pieces are the network packets.

Some network applications distribute the network packets in clear text. These packets are not encrypted and can be processed and read by a packet sniffer which can be implemented in either software or hardware. The packet sniffer is normally used for network fault analysis. It can pick the packets off the network and process them for data analysis.

The packet sniffer software will capture the packets received on a physical network wire. There are freeware and shareware packet sniffers that allow any user to get into a system without any knowledge of the inner workings of the system.

In a networked database, a packet sniffer can be used to get information that is queried from the database, along with user account names and the passwords used. This can be a problem since users often reuse their log-in names and passwords across multiple applications.

Other packet sniffer information includes the topology of the network, what computers run specific services, how many computers are on the network and which computers have access to others.

Password Attacks

Password attacks occur when there are repeated attempts to identify a user account or password. These attacks take place from brute force, repeated attempts which are often automated. Other attacks can come from Trojan horse programs, IP spoofing, and packet sniffers.

Once an attacker gains access to the network, they could modify the routing tables for the network. Then, all network packets can be routed to the attacker before they are transmitted to their final destination. This allows the attacker to monitor all network traffic and become a middle man.

A middle-man attack can come from someone who is working for the Internet service provider (ISP). This person could gain access to the network packets transferred between the network and any other network using packet sniffers and routing or transport protocol software.

These attacks could result in the theft of information, hijacking of internal network resources, denial of services, corruption of transmitted data, introduction of new or false information into network sessions, interruption of control and the degradation of network performance.

IP Spoofing

An IP-spoofing attack can occur when an attacker outside the network pretends to be a trusted computer on the system by using an IP address that is within the range of IP addresses for the network. The attacker may also use an authorized external IP address to provide access to specific resources on the network.

IP-spoofing attacks are limited to the injection of data or commands into an existing stream of data passed between a client and server application or a peer-to-peer network connection. For bidirectional communication to exist, the attacker must change all routing tables to point to the spoofed IP address.

An attacker can also emulate one of the internal users in the organization. This could include e-mail messages that appear to be official. These attacks are easier when an attacker has a user account and password. But, they are still possible by combining spoofing with a knowledge of messaging protocols.

Denial-of-service Attacks

Denial-of-service attacks concentrate on making a service unavailable. This can be done exhausting or slamming a server on the network. Denial-of-service attacks can use Internet protocols, such as TCP and the Internet Control Message Protocol (ICMP).

These attacks probe a weakness in the system. Flooding the network with undesired network packets is one technique that is used. Another technique is to leave false information about the status of network resources.

Application and Trojan Horse Attacks

Application attacks use some weakness in the software that is found on servers, including FTP. This allows the attacker to gain access with the permission of the account running the application.

Trojan horse program attacks use a program that the attacker substitutes for another program. The substitute program can furnish the functions of the normal program. But, it also may monitor log-in attempts and seize user account and password information.

One type of Trojan horse program displays a screen, banner, or prompt that looks like a valid log-in sequence. The program processes the information that the user types in and e-mails it to the attacker.

Securing the Network

Network security can include firewalls, proxy servers and data encryption. Proxy servers and firewalls act as traffic cops. They regulate who gets in and out of the network.

A firewall monitors the traffic crossing network perimeters and sets restrictions according to the security policy. Perimeter routers can also be used. These can be located at a network boundary. Firewalls commonly separate internal and external networks. The firewall server acts as the gateway for all communications between trusted networks and untrusted and unknown networks.

The least secure network is the outermost perimeter network of the system (Figure 7-6). This area of the network is the easiest area to gain access to, making it the most frequently attacked.

Passwords

The shorter the password, the easier it is to guess and test. Passwords should be at least eight characters long. The maximum number of

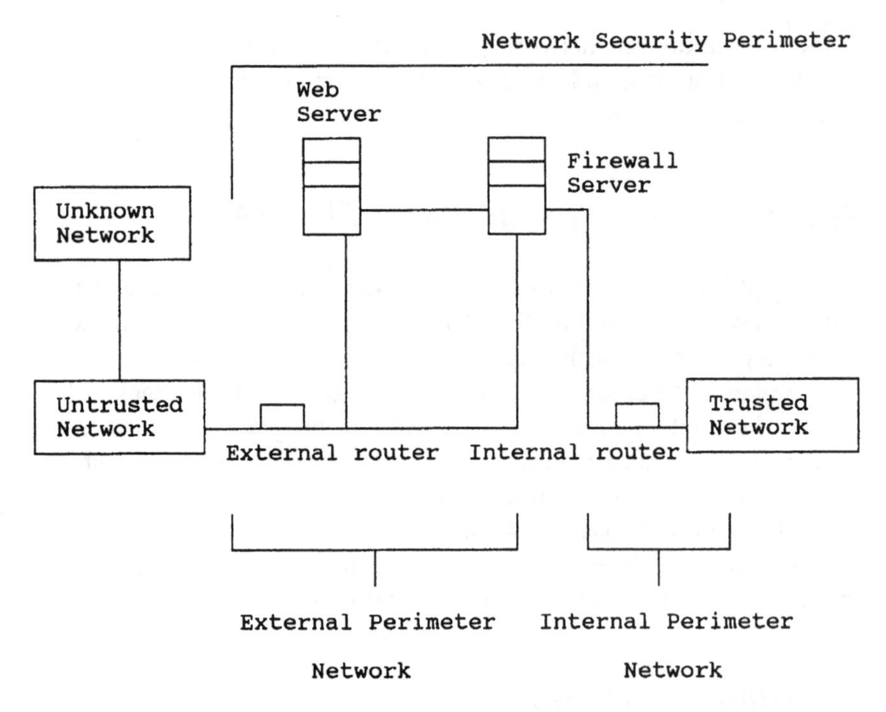

Figure 7-6. Perimeter networks that need to be protected.

characters you can type in the log-on box will make it harder for hackers, but also harder for users to log on. Place restrictions on all user-defined passwords. Do not use words for passwords that contain the birth date, middle name, and/or spouse's and children's' names, since this makes it easier for hackers to obtain access to the password.

Also, do not use any common word that can be found in a dictionary. A common hacker technique is to run lists of common words such as names of famous places or people. Use more complicated rather than simple passwords. The later Windows NT versions have a Service Pack with features that improve the security of the network passwords. It requires the user passwords to be at least six characters long. They must also contain characters from at least three of the following four groups:

- English uppercase letters (A-Z),
- English lowercase letters (a-z),
- Arabic numerals (0-9) and
- Punctuation or other special characters.

Many Internet sites require you to enter a user name and password in order to access information. You should never use your Windows NT log-on password to sign on to a Web site. There is the chance that someone could intercept the password and try it to log onto the system.

Log-in and Registry Security

There is a Microsoft utility called TweakUL. This utility stores the user name and password values in the Windows Registry in an unencrypted format. Hackers can access the Registry. If the system is set up for automatic log-on, you can still protect the log-in information by editing the permissions associated with the segment of the Registry where the information is automatically stored.

You open the Windows NT registry Editor and choose Permissions from the security dropdown menu. Then change the default permission for the Everyone group from Read to Special Access. Then, deselect the Query Value check box. Now, when someone other than an administrator tries to access the key, it will appear grayed out.

Windows NT can clear the user name on the log-on screen. NT by default displays the name of the last person who logged onto the system. This can be a security threat, since a hacker may guess a user's password based on the account name or the log-in environment. You can clear the previous user name by adding a new string value called

DontDisplayLastUserName

Security is improved if there are only READ capabilities on the Registry keys. The default setting is WRITE on some of the most important keys. These WRITE privileges allow a user to gain access to the server. This can be prevented by changing the permissions on these keys to READ. Use the Registry Editor to change the security/permissions on these keys.

When a user starts an NT workstation, it automatically creates some shared resources. These include a share to the root directory of the hard drive that remains invisible during browsing. To prevent this automatic resource sharing you can edit the NT Registry settings.

Windows NT holds data in swap files and temp files. In order to speed access and improve performance, pages of data are swapped in and out as needed. When the system shuts down, the swap file stays on the hard drive. An intruder could view data stored in the swap file. The NT system can be configured to clear the swap file every time the system shuts down.

This can be done by using the NT Registry Editor. You go to

HKEY_LOCAL_MACHINE

and add

ClearPageFileAtShutdown

This will clear all pages except those that are still active during shutdown.

One of the common paths a hacker uses to gain access to a network is through services running on the server. These include the ClipBook Server, Plug-&-Play and Net Log-on. Consider shutting down those that are not needed.

Smart Networks

Smart control and advanced networks define the future of control technology. The goal in the future will be to integrate plant-wide controls with the monitoring of energy used for heating and ventilation.

Fan rooms are having their older pneumatic controls replaced with modern data acquisition and control. Field devices are used as input/output devices to the PC in each fan room substation. Windows NT is used in a PC-based control system with fieldbus communications. Benefits include lower installed costs with less wiring and true distributed intelligence. Communications are improved as well as diagnostics between devices. Incorporating all control, operator interfaces and data acquisition in a single system allows a leaner overall operation. Hardware savings can reach 25%.

PC-based software like Paradym-31 provides ladder logic, sequential function charts, and function-block diagrams. User-defined function blocks allow users to customize the HVAC application using software tool sets.

PC-based control will be used for more energy saving applications. It can provide control capabilities usually missing in older classical methods.

Pure PC-based control will grow while blending traditional forms of control like programmable logic control with PCs will decrease as the range of PC software expands to cover more diverse applications.

PC-based control runs on a standard operating system such as Windows NT. Off-the-shelf application programs continue to improve. Control systems can be built on Windows with real-time extensions.

PC-based control offer improved options in programming software. Systems can utilize popular languages such as Visual Basic, Visual C++, C, or even ladder logic. Applications programmed in different languages can reside simultaneously on the same system.

The methods used to simplify program control continue to grow and improve. These include sequential function charts, function-block diagrams, structured text and flowcharts. A system may use a combination of several languages and program control methods.

In PC-based control, the information sharing between programs continues to grow and become more simplified. Tools like ActiveX controls allow small, specialized software components to be created that can be reused to develop customized applications.

I/O independence is another characteristic of most PC-based engines. It allows users to choose the I/O and network solution best suited for the application instead of using a proprietary solution. This is being fueled by open, multisupplier network technologies such as DeviceNet, ControlNet, and Ethernet.

Open Systems Growth

Today, there is a trend towards open computer systems for commercial and industrial control. These applications include boiler and chiller controls, heating and air conditioning and fire alarm systems. In these control and monitoring systems, a critical element is the method of communication that links the individual components into an organized system.

Most modern control components are capable of connection to some type of communication network. There are many choices available as these open control systems compete for attention.

The elements in these systems need to pass information among themselves in order to execute their assigned tasks. An example is an outside temperature sensor for hot water reset, which signals a boiler controller to produce hot water at the correct temperature to satisfy a building's heating load. The use of installed equipment from multiple vendors can be expensive to interface and the installation of control equipment from another manufacturer requires new skills.

As products improve, the life cycle for these systems is becoming shorter. Parts and service for older systems can cause costly delays and it may become more economical to install newer systems that can provide more efficient performance along with modular design for easier maintenance.

A proprietary control system usually requires proprietary hardware from the original manufacturer. The move to interface computer control systems from different manufacturers is driving the growth of open systems for commercial and industrial control.

The open system notion allows the connection of control system components from different manufacturers to operate as an integrated system in a mode which is transparent to the user. The system components also become more competitive with improved availability.

The open systems concept has been demonstrated in IBM personal computer systems. These computers are based on open specifications and any manufacturer can build systems around them. This results in system interoperability in a more wide open market.

Open System Issues

As open systems come into widespread use, several issues need to be evaluated. The protocol functions supported need to include reading digital and analog values, changing setpoints and schedules, and downloading data.

Operator workstation functionality may be limited when communicating with the open protocol equipment. Ideally, conformance with interoperability should be plug-and-play.

The functionality of the operator workstations are important since system configuration, routine engineering and maintenance functions are not usually performed at a single location.

Several trade-offs are involved in using open systems. These include performance limitations, system integration, maintenance, and service issues.

The mixing of open system standard protocol capabilities with proprietary system functionality leads to increased costs and creates potential purchasing and implementation problems.

Detailed functions and services must be specified since many standard protocols do not provide the same capabilities. These include monitoring, alarm, setpoint changes and scheduling functions.

The definition of system integration responsibilities is important when using equipment from different manufacturers. These responsibilities include integrating new equipment and the database at the operator workstations to incorporate new requirements. Changes in one manufacturer's hardware or software may affect other portions of the system.

The definition of factory and field testing is also important. This includes the acceptance criteria, based on individual component and integrated system functionality. This functionality must be tested at both levels. It is not enough to test the monitoring and control functions supported by the open systems protocol from the workstation to each field panel. The programming functions required at each field panel location must also be tested.

Maintenance is a critical issue for the successful operation of the system. System maintenance refers to both hardware and software. Hardware maintenance is simplified because of built-in equipment diagnostics and plug-in replacement boards. Software maintenance can be more demanding since there may be multiple databases and application programs to maintain as well as several different programming languages. Electrical and mechanical system maintenance programs can be used for preventive maintenance and trend recordkeeping.

System Hardware

Software is only part of the solution, the hardware on which the software runs is also important. The PC system can be an industrial computer, commercial-grade computer, or open type of controller. The hardware usually depends on the environment in which it will run. Industrial computers are available with shock-mounted hard drives, ball bearing and filter cooling fans, internal temperature sensors, redundant hard drives, system control monitor cards and uninterruptible power sources.

Windows NT/2000 Changes

Windows NT Version 4 is being released as Windows 2000. It has added support for DirectX, which is Microsoft's multimedia platform, changed its driver architecture to be compatible with Windows 98 plug-and-play, and made a number of other changes. With the introduction of Windows 2000, Microsoft has made some strict rules regarding what can go in the SYSTEM directory of Windows NT.

For system users, these changes have been largely a plus, adding capability and compatibility. Most beta testing indicates little or no difficulties in migrating from Windows NT 4 to Windows 2000. Preliminary testing of existing applications in Windows 2000 Professional and Windows 2000 Server has resulted in only minor setup problems.

Several companies such as Intellution, Wonderware and SoftPLC have developed programs that run on a PC that emulate a PLC. These programs use ladder logic and allow the PC to function like a PLC.

There is a development platform and a run-time platform, and in some cases an operator interface front end. Most programs such as Paradym-31 from Intellution have a self-documenting capability and can go from ladder logic to other programming techniques.

Software Suites

The trend in control automation is towards tightly integrated software suites. The access of diagnostic information is a major part of the control management and configuration software. OPC enabled automation suites may include the supervision, operator interface, reports, diagnostics, calibration and maintenance. The configuration package may include the construction of control strategies and simulation.

OPC is bringing major changes in the way diagnostics are managed. Fieldbus devices provide a driver for OPC, just as printers use a driver for Windows.

Companies that previously only dealt with software may now sell it preinstalled in a PC, with interfaces as a console product. Users then plug in the field devices without the setup problems of cards, drivers and interrupt conflicts.

Users can ensure that they get devices with the diagnostics they need to reduce control variability and downtime. Fieldbus devices are available to provide the diagnostics needed. In the display end, efficient interfaces and software with extensive template facilities should be used.

The concept of an industrial control network is to link all devices or stations with one cable. No matter how many devices you are connecting, you just add the physical cable which is needed to connect to the furthest station. Troubleshooting becomes much easier since you can plug and unplug the device from the network. This virtually implies a plug-and-play ability of the devices.

The best of buses for one industrial automation application may be the worse of buses for another. Wide-ranging requirements have spurred the creation of specialized networks.

The move to open digital networks marks a significant change in the architecture of control systems. As with any major technological change, much contention surrounds the choice of technologies, methods, and architectures.

The IEEE-488 standard has been the dominant communications interface for electronic instrumentation for more than 20 years. IEEE-488 has evolved to include more software standardization and increased transfer rates. It continues to dominate test-and-measurement applications.

Several major technological shifts have occurred in control systems over the last half century. Process control has shifted from manual systems to pneumatics, followed by electronic control systems based on the 4- to 2-mA instrumentation standard. Many distributed control systems still use 4- to 2-mA instrumentation wired to I/O subsystems, which in turn are connected to larger systems by proprietary digital networks.

The most significant move in the early 1970s was from hardwired relay panels to programmable logic controllers (PLCs). These PLC-based systems directly wired the analog and digital signals to I/O subsystems, which were connected to the PLC using vendor-specific digital networks.

Devices that include both digital-communication and sensing capabilities distinguish today's digital networks. A common network that directly connects the control system's measurement and control devices replaces point-to-point wiring to centralized I/O subsystems. Digital networks provide several benefits. One of their most basic characteristics is the increased information available from devices that communicate digitally.

Open Systems

The detailed concept of open systems for computers has developed from the Open System Interconnection (OSI) reference model for computer communication standards. The OSI model has seven layers of communication protocol and each of the layers has its own function with defined interfaces to higher and lower layers.

These layers are used with programming object models to define the meaning and format of data. This can provide a high level of multivendor interoperability.

In an ideal open systems architecture, devices communicate with each other using a standard protocol as a native language, without translation. The ideal system would have interoperability at all levels.

Some so-called open systems interface workstations and control system components from one manufacturer with other manufacturers' components using gateway hardware or software on the network segments. The native language may be a proprietary protocol in this system.

The system is forced to be compatible, using translation in the form of the gateways implemented in hardware or software. The devices in a communication segment must be from the same manufacturer. These devices may include terminal units, chillers, fire alarm panels, security access controls, generator controls, fume hoods and programmable logical controllers.

In a true open system, the communications between workstations and field devices take place on a common bus using non-proprietary protocols. The field devices may be setup internally to perform the gateway functions, or they can use external gateways to communicate with other manufacturers' equipment using a proprietary protocol. Protocols provide for the data exchange of a defined set of programming objects and services.

In Europe, open systems development has produced Profibus and FND. In the United States, the Distributed Network Protocol (DNP) and Utility Communications Architecture (UCA) were developed for industrial and utility Supervisory Control and Data Acquisition (SCADA).

Data communication protocols for building automation and control networks include BACnet, sponsored by ASHRAE, and LonTalk, sponsored by the Echelon Corp. The BACnet protocol has been adopted as an ANSI/ASHRAE standard.

Several fieldbus technologies are already de-facto standards and competing for market share in HVAC controls. Among building automation trends this will be one of the most significant changes in the near future.

Fieldbus Diagnostics

Information is the currency of the information age and fieldbus-based devices provide control variables as well as lots of noncontrol information. Data about equipment exceeding operating conditions and other abnormal behavior can be used for predictive maintenance scheduling, which minimizes unplanned downtime. Detailed device self diagnostics and automatic failure alarms can be used to determine if a problem is due to the sensor or without involving field visits.

Field device management software can complement the basic control function, reducing configuration, calibration, and maintenance costs.

The link between control systems and business systems can include resource utilization planning with the help of diagnostic information from field devices. Fieldbus provides a means of communicating the information from field devices to a host system. Emerging standards will define interfaces from the control system to the business level.

Other technology, like the specification for OLE for Process Control (OPC), make it easier to bring this information into the host system.

Unplanned shutdowns due to failures can be avoided if users can schedule controlled shutdowns when a number of devices need repair. Spare parts can be ordered in advance, avoiding delays. Good diagnostics can help avoid accidents.

Field devices in HVAC systems include sensors and positioners. Temperature and pressure sensors are dedicated to providing the processor with information on the state of each device and ambient conditions.

For positioners the possibilities for diagnostics are larger because positioners control their outputs and sense the feedback. This allows positioners to perform tests to determine, for example, if the function of a valve or damper is correct with respect to the positioner output.

Self diagnostics of devices can notify the operator of failures and problems, allowing maintenance to pinpoint errors. Sensor and calibration errors are reported automatically without the need for manual interrogation. Devices can provide automatic safety actions to prevent dangerous conditions. Diagnostics may be checked directly from the operator console or from hand-held terminals.

Continuous device self diagnostics can be reported as events as well as configuration diagnostics. User-invoked diagnostics tests can be used to check damper and valve operation.

Modern HVAC controls must maintain accurate and reliable measurements to achieve lower operational costs. Fouling of sensors may be detected as well as faulty or poor wiring. Long-term drifts or shifts due to conditions or aging can be detected if there is a comparison against an external reference providing calibration.

Positioner diagnostics can include diagnostics of actuators and valves. Valve positioner diagnostic information can include valve stem friction and other parameters. Statistics on cycle count and time in operation can be recorded to estimate wear and tear and predict failures. Other data from the positioner can be used to determine positioner and valve characteristics, such as valve position travel time.

Stored device information includes tag and serial numbers, characteristics and date installed. Storing the information in searchable and sortable databases for easy access reduces errors and makes calibration and preventive maintenance less costly while ensuring it is done correctly and on a periodic basis. The result is better measurement and control.

Fieldbus places new requirements on host operator interface software. Older distributed control systems (DCS) and PC-based software was designed in an era when field devices provided only a single value. With fieldbus, there is a huge increase in diagnostic and other data.

Some of these are continuously updated and others are updated as alert reports. Others are static and updated only when they change.

The information had to be entered for each parameter can include node address, data file and memory address. Data type and display and alarm information may also be applicable to some parameters.

Since fieldbus devices are standardized and based on user-defined tags, the user can specify the tag of a function block or device and all the diagnostics and pieces of information associated with that tag are automatically configured in detail.

Automated configuration generation reduce the number of hours to make a configuration and users will get more data easier and in less time.

Efficient data access includes logical groupings of associated parameters that displayed together. It allows the access of information as a single communication, rather than many.

Devices in the field detect any alarms and report them only when they occur and disappear. This is far more efficient than for the host to poll the device for the variable a few times per second and detect the alarm in the host. Field devices can also help the host in trend recording by taking real-time samples and passing them to the host in a single communication.

An important function of fieldbus is the mechanism for accessing static parameters only if they are changed. Field devices automatically notify the host when changes are made to them. Other functions traditionally done in the host that may be performed by the field devices include setpoint generation and logic functions. Field devices can also take over controller functions like PID and computation from the DCS.

The increased amount of data associated with diagnostics and other functions in fieldbus requires software that must make use of the alarms and trending.

OPC is addressing this capability. The RS-232 port on a PC is a bottleneck and is not fast enough to handle several fieldbuses. Special multiport fieldbus cards with direct memory access and dedicated onboard CPU for data access can provide a bridge between network segments and provide communication management. This CPU operates independently of the PC and can even perform some advanced control and computation functions.

Information can be distilled into a simple, organized form. Hosts can collect as much data as possible and then select what is important.

Host displays normally show overview screens with overall states and not too much detail. The operator can then zoom in on the details in a hierarchical manner. When a parameter deviates from its normal value, the operator must be alerted.

Besides looking at individual loops, there must be screens showing device and function-block status and the state of the communication subsystem. There must also be access to specific diagnostic information about the loop or device to determine the nature of the fault. The software should store and be able to recall the values, status, and alarms just before the failure occurred.

References

Agarival, Anant, "Raw Computation," *Scientific American*, Vol. 281 Issue 2, August 1999, pp. 60-61.

Berge, Jonas, "Fieldbus Advances Diagnostics," *InTech*, Vol. 45 No. 4, April 1998, pp. 52-56.

Byres, Eric, "Ethernet to Link Automation Hierarchy," *InTech*, Vol. 46 No. 6, pp. 44-47.

Evans, Christopher, *The Micro Millennium*, Viking Press: New York, 1980.

Gearhart, Michael A., "Open Communication's Protocols Emerging," *Access Control & Security Systems Integration*, Vol. 7 No. 10, July 1, 1997., pp. 1-5.

Graham, Jamie, "Beef Up the Security of Your Plant Network's Internet Connection," *Instrumentation and Control Systems*, Vol. 72 No. 8, August 1999, pp. 39-47.

Hogan, Mike, "The PC of Tomorrow," *PC World*, Vol. 16 No. 1, January 1998, p. 138.

Liptak, Bela A., Editor-in-chief, *Instrument Engineers' Handbook*, 3rd Edition, Chilton Book Company: Radnor, PA, 1995.

Naugle, Matthew G., *Local Area Networking*, McGraw-Hill, Inc.: New York, 1991.

Nemzow, Martin A., *The Ethernet Management Guide: Keeping the Link*, McGraw-Hill, Inc.: New York, 1992.

Saffo, Paul, "Future Software: Where do You Want to Go Tomorrow?," *PC World*, Vol. 16 No. 1, January 1999, p. 140.

Chapter 8

Open Systems and Device Networks

In today's energy management climate, increasing costs and the need for improved efficiency of building systems forces owners to take a hard look at their building automation systems.

Many industries must comply with new environmental and safety requirements. In such a business environment, less of the required data are control specific. More information of other types is needed to maintain, diagnose and modify the control system.

Standard digital networks have additional benefits. Reduced wiring lowers installation and maintenance costs. Open systems allows users to select the right equipment for the job, regardless of the system manufacturer. The networks also have intelligent instrumentation, in which better performing devices provide functions such as advanced diagnostics.

The networks offer distributed control with intelligent devices that provide the flexibility to apply control centrally or at local processing points for improved performance and reliability. Those available range from low-level systems that gather data from simple discrete sensors to more sophisticated networks in which intelligent field devices implement distributed control.

Open and Standard Protocols

There is some difference between an open protocol and a standard protocol. Open means that anyone is free to use the protocol. Standard means the document that describes the protocol has been through a process and been reviewed by a variety of interests which may include users, vendors and supporting agencies.

A standard protocol for building control can have great benefits. Standard protocols for open systems allow interoperability of different systems. Some benefits of open protocols cannot be realized without industry standard protocols.

A standard protocol provides vendor independence since it allows the interoperability of different vendors' systems and components. Interoperability allows competition which tends to regulate cost and prices. It is always best to make use of existing standards to the greatest extent possible. Most vendors and suppliers know that interoperability standards are important.

Integrating equipment and controls from different companies to create a reliable control system has been one of the toughest tasks in building automation. Non-homogeneous systems increasingly need to share information and it is important to interconnect more easily.

HVAC controls are available for many open systems. The shift from proprietary to open control systems is similar to the shift that occurred from proprietary mini-computers to generic, IBM-standard personal computers based on open specifications.

In the computer industry before 1980, it was a vertical market since each vendor provided all of the major components of the system. Once a specific vendor was selected by the customer, service, future additions and changes to the system were generally available only from that vendor. The customer was locked in.

Today, computers are sold through the mail. The market is no longer vertical. It is horizontal. Customers can mix and match if they desire. They have choice and flexibility. The computer industry could not have shifted without a set of rules. Hewlett Packard laser printers can communicate with Intel chips and are directed by Microsoft operating systems.

The absence of a set of rules kept the market vertical. As the rules evolved, a new organization took shape. This shift provided new roles for computer manufacturers. Some continued to build complete systems and still do. Over time, others focused on what they did best. The industry was essentially transformed, with a proliferation of new products.

The building control industry is beginning to follow this pattern. Industry sectors like actuation are developing with companies such as Belimo and others specializing in this area. Performance contracting is another growing specialty.

An example of an open system occurs in computer-aided design

(CAD), where DGS, and DXF file interchange protocols allow the AutoCad and Microstation CAD systems from different software vendors to exchange graphic files.

Building Automation Trends

Open control systems using BACnet and LonTalk are becoming more common. Product interoperability depends on the BACnet committee's documents on minimum BACnet performance requirements for different types of equipment and the developers of LonTalk, under the LonMark trade association, developing similar performance requirements for different types of mechanical and electrical systems.

Open system product offerings include building automation systems, DDC controls, security systems and fire alarm systems. Some manufacturers support one or both of the leading commercial protocols, BACnet and LonTalk.

There are over a dozen open system protocols for building control systems. One of these is BACnet, which was developed under the auspices of the American Society of Heating, Refrigerating and Air Conditioning Engineers (ASHRAE).

BACnet

When the American Society of Heating Ventilation and Air Conditioning completed its 8-year effort to develop the common protocol called BACnet the goal was to establish interoperability between DCS equipment manufacturers. Echelon is the standard at the sensor bus level.

The BACnet (Building Automation Control network) protocol allows building automation equipment from different manufacturers a minimum level of interoperability. ASHRAE began developing BACnet in 1987. It was published in 1995 as ASHRAE Standard 1355-1995. ASHRAE'S BACnet standard holds great promise for the HVAC industry. Most control manufacturers have announced their support for the standard, and many are providing BACnet products today.

BACnet is a software solution to integrating different products from different manufacturers. It consists of a standardized set of messages between field units and the host. BACnet is also a recognized standard of the American National Standards Institute (ANSI) which has a BACnet device compliance testing facility in Gaithersburg, MD.

BACnet was developed as a voluntary consensus standard as ASHRAE Standard 135. There are three main parts to the BACnet standard. There is a specification for local area network technologies. A local area network (LAN) provides a way of moving information among computers. Another part of the BACnet standard provides a way to interconnect multiple LANs into a wider area network. A third part of BACnet specifies a standard communications format that can be used by different manufacturers to allow their building control systems to work together as independent but cooperating systems.

BACnet Application

BACnet applications include integrating fire detection, protection, and suppression systems with HVAC systems and increased communications between building control systems and utilities for energy management and utility load management. BACnet can provide better energy management, building performance monitoring, and enhanced HVAC diagnostic capabilities.

A consortium of control manufacturers organized by the National Institute of Standards and Technology (NIST) has been testing manufacturers' equipment for interoperability.

ASHRAE has been developing an addendum to BACnet that would define conformance tests. These tests could provide an industry-wide certification program that would ensure the compatibility of BACnet products made by different vendors.

Manufacturers are applying BACnet for island-to-island, multi-vendor system integration. Communicating through BACnet, these islands of control allow monitoring and command capabilities from a single PC workstation. BACnet can also be a vehicle for combining multi-vendor campus sites into a single intercommunicating system and moving existing systems into an open environment.

When a building owner buys a building control system, the owner is effectively locked in. Once a specific vendor is selected by the customer, ongoing service, future additions, and changes to the system are available only from that vendor. Building owners that find themselves limited to a single supplier for service and upgrades have higher costs and fewer options.

BACnet provides an alternative since there are a set of rules that allow the major components of a system, even if they are manufactured by different vendors, to function together. They can interoperate. This interoperability promises to provide competitive pricing, more choices for products and service, and the ability to even operate two different systems with a single computer station.

BACnet is well documented and robust in design. It has the features needed and has headroom for the future. Manufacturers that see BACnet as an opportunity will adopt it.

The complete BACnet standard covers about seven different data link layer models. But, a BACnet compliant product on an ARCNET level and a BACnet compliant product on a sensor level does not establish interoperability. If BACnet is not used internally in some new products, then a gateway will be required and that gateway can act as a firewall.

A firewall might be used by an equipment manufacturer of chillers and boilers, to protect the equipment's internal safety logic and critical control algorithms. If this logic is compromised, it can create a warranty or liability issue.

At minimal compliance, it does not make much difference what kind of system or what communications protocols are used in collecting data or in issuing field commands. What is important, is what data are gathered from any given piece of equipment, and how that data are structured. Using this strategy and basic network tools, information gathered from multiple field systems can be presented as a single system, to multiple users.

One problem with a BACnet minimum level of compliance, is that it is not easy to make programming changes. In most cases, programming must be done at the controller connected to the field equipment.

Full compliance allows programming as well as monitoring and control functions at the user level. Most manufacturers have offered BACnet compatibility at the minimum compliance level.

Conformance Classifications

The BACnet standard defines conformance classes which are groups of functions and capabilities needed to allow specific levels in BACnet communications. The protocol's hierarchy of conformance classes means that the goals of the project become important in writing BACnet specifications. There is an effort to make BACnet easier to specify and to simplify the BACnet specification process. The committee charged with the maintenance of the BACnet standard (ASHRAE SSPC-135) has been redefining how to specify BACnet.

The BACnet standard has detailed information on how to implement the protocol. Clause 22 is devoted to the specification of BACnet systems. It involves the concepts of BACnet Conformance Classes and Functional Groups. The Conformance Classes range from 1 to 6. The standard states that they are not hierarchical and that Class 1 is not better than Class 6.

Product Implementation Conformance Statements (PICS) from vendors determine which products can be used together. BACnet interoperable systems should be able to provide single-seat operation where a central PC operates buildings with control systems furnished by several suppliers. It should allow equipment integration including chillers, rooftop units, fire alarm, and lighting control systems.

There will be an expanded availability of intelligent devices that are supplied with intelligent electronic panels or interfaces that communicate with the host. These will integrate lighting, security, access, and fire alarm controls into one system, so the building can be run from a single operator interface.

BACnet is not a plug-and-play system nor are the other interoperable options available today. It is not possible to remove a VAV box controller and replace it with one from another supplier without performing additional setup and programming. These functions will continue to be achieved in a unique manner by each supplier.

Interoperable systems allow the following tasks to be accomplished in a common format:

- viewing data on a PC (menus, graphics, reports),
- changing setpoints,
- sharing data between controllers,
- viewing alarms for points that are out of a defined range,
- setting up and modifying schedules, and
- sampling data at regular intervals and retrieving it for viewing or storage.

The first three functions will be able to be supported by even the most basic BACnet device. This data exchange capability should be supported by suppliers of BACnet systems, as well as suppliers of BACnet gateways. A PC workstation performs data exchange by checking the information that is contained in BACnet controllers and changing setpoints and other values on these controllers. This is done with a series of BACnet functions or services called Read Property and Write Property.

This interoperable capability allows a workstation to display temperature, pressures, and other values. It also allows the operator to change setpoints, and store data for historical trending.

BACnet supports different network methods including Ethernet. Read data are used for viewing on graphics and write are used for changing setpoints. In BACnet the workstation needs to use the Read (Initiate) and Write (Execute) Services in Clauses 15.5 and 15.8 of the ASHRAE Standard 135-95. It uses these services to communicate with BACnet objects in the internetwork.

These objects are also known as points and to perform a function, a point or object list is used to show how an object is related to a service in the BACnet internetwork.

Legacy systems need a gateway to connect the BACnet and the legacy system. The gateway will transfer alarm, trend, schedule, and point data between BACnet and the legacy system.

BACnet Interoperable Functions

Data exchange between devices such as PC workstations and building controllers is the most basic interoperable function. It involves the viewing or reading of data as well as making changes or writing to this data.

In BACnet, data are modeled as objects that have properties. An alarm and event function allows the exchange of alarm information. This function can be used to trigger a control action, or an operator override. A set of functions allows the editing and creation of schedules on a BACnet PC workstation.

A network management function provides the ability to manage the devices on the network. This includes monitoring for a loss of communication and coordinating the time settings of the clocks in each controller.

Interoperability will be realized only if BACnet is supported widely by vendors. Most of the major controls vendors are introducing or expanding their BACnet-based product lines. Many BACnet product lines do not use BAController level and at the highest levels of the system they use a BACnet conformance class of only about three or four.

LonMark

In May 1994, a group of companies saw the need for establishing an interoperability standard for control networks and formed the LonMark Interoperability Association. The association has over 200 members and promotes LonMark products and systems as open, interoperable control solutions. It sets the design guidelines for interoperable devices based on LonWorks and verifies products that meet the guidelines for interoperability.

Products that have been verified to conform to LonMark interoperability guidelines carry the LonMark logo, which indicates that they are designed to interoperate across a LonWorks network. LonMark task groups include elevator, HVAC, lighting, refrigeration, fire, industrial, network tools, security and system integration.

Interoperability means the easy integration of products from multiple vendors without the need for custom hardware or software. With interoperable control devices, end-users and network integrators can purchase devices off the shelf from different manufacturers. Plugging in these interoperable devices allows easy expansion of control networks.

LonWorks Control

LonWorks provides an open, off-the-shelf networking technology platform for designing and implementing interoperable control networks. LonWorks represents a complete hardware and software solution for control network communication. It does more than integrate devices and control systems. It offers a single, cohesive, interoperational control system that does not depend on one proprietary vendor or require costly customization.

If the lighting and heating, ventilating, and air-conditioning systems have LonWorks technology embedded in them, the two can be connected. Now, in addition to shutting down lights, the occupancy sensors can send a signal to the HVAC system to adjust heating or cooling levels as well.

The LonWorks system was designed by Echelon to use a proprietary integrated circuit called the neuron chip. A special communications protocol is used with these chips which are manufactured by several companies, including Motorola and Toshiba. LonWorks allows a seamless connection between systems that are designed using the same parameters.

The neuron chip is designed to communicate with field variables, run process control loops and communicate with other neuron chips. The chip can be used for monitoring and control at different levels in the system. Some manufacturers even use the neuron chip in the main controller of their EMCS system. While these systems use LonWorks, they are not true LonWorks systems.

Open systems like LonWorks also allow customization. Open, interoperable products and architectures allow the overall infrastructure and system to fit the needs of the end-user and the facilities manager.

Once the infrastructure is put in the building owner has a fully integrated, intelligent unit. Applications can be added as needs change and grow. As new sensors and applications are developed, you can just change components.

The hardware platform is based on the Neuron chipset, which provides I/O processing and data communications. Software is based on the LonTalk communication protocol, which provides a set of standardized messages and a development tool for applications.

The original implementation of LonWorks required a licensing agreement from Echelon. Since that time, Echelon has allowed users to port LonTalk protocol to a number of different hardware platforms that provide increased performance over the 8-bit Neuron processor.

BACnet and LonMark

Manufacturers are delivering products that are compatible with both BACnet or LonMark. While BACnet and LonMark have their own benefits, manufacturers are finding them to be complementary. The complementary nature of the two standards allows different building control systems to communicate with each other and exchange information.

BACnet and LonWorks are not interchangeable since each has its own standardized set of messages. However BACnet messages can be transmitted across a LonWorks communication backbone. BACnet recognizes LonWorks as a physical medium only, but messages between LonWorks and BACnet are not interchangeable. LonMark is efficient for low-speed, field-device communication, but does not support the high-speed exchange of data on networks like Ethernet as well as BACnet.

LonMark is being applied for peer-to-peer, distributed control solutions. This type of control uses a network architecture, with a workstation, field panels, controllers, and individual devices. LonMark allows intelligent devices to handle information and pass it on to other devices. LonMark is a cost-effective technology for interoperability at the controls level.

LonTalk Protocol

An open technology like the LonTalk protocol is embedded on a chip. Manufacturers of different products include the chip as a standard part of their unit.

The protocol provides a set of rules for the connection and makes it cost-effective and forces the implementations to be consistent. The protocol chip makes it cost-effective for the manufacturers and makes every implementation exactly the same so that there are not any different interpretations of a specification.

Neuron-based, open-standard protocol devices are used by electrical manufacturers and system control vendors. The equipment remains autonomous and fault-tolerant as centrally managed programs become distributed throughout the LonWorks network and each Neuron device learns its attached equipment's role and relation to other equipment.

Electronic Systems USA is a building automation products and services company that produces products with LonWorks technology, such as the NexSys building automation system. LonWorks allowed Electronic Systems to get to the marketplace quicker because all the protocols are set. They did not have to develop a new system.

The neuron chip is programmed with a version of the C language called neuron C. It can also be programmed using drag and drop software from Dayton General Systems which allows the programming to be done much faster.

LonMaker for Windows is a design tool that makes it easy for control and facilities managers to design and maintain LonWorks networks for heating, ventilation and air conditioning systems. Formerly incompatible data formats and network databases are being bridged by LonWorks technology. These systems automate equipment reconfigurations, improve efficiency, and reduce personnel and utilities costs.

LonWorks networks are built using the LonMaker for Windows. The Visio Technical drawing package makes up the LonMaker user interface. Dynamic Data Exchange applications, such as Microsoft's Visual Basic or Excel, are also supported.

Cisco Systems and Echelon have partnered to create a standard for merging information processing IP and LonWorks networks more fully. As the dominant router supplier, Cisco sees a future in augmenting data networks with LonWorks systems. Echelon is also working with Sun Microsystems to define standards for integrating Java applications with LonWorks devices using Sun's Jini technology. Microsoft is also promoting the merger of electrical devices with data networks in a more high-level way, by advancing its Universal plug-and-play protocol.

DeviceNet

DeviceNet is a CAN-based bus developed by Allen-Bradley. It is a low-level network that connects industrial devices, such as sensors and actuators, to higher level devices such as variable-frequency drives and operator interfaces. DeviceNet is designed as a low cost peer-to-peer network. Although this open industrial control networking system was first sanctioned by Allen-Bradley, it is now sustained by an independent organization, called the Open DeviceNet Vendor's Association (ODVA). The Open DeviceNet Vendor's Association provides free technical support and has DeviceNet experts around the world to answer technical questions. DeviceNet uses the Controller Area Network (CAN) as a backbone and follows the ISO/OSI 7-Layer model.

This 8-byte fieldbus system is designed for industrial I/O control of sensors, switches and motor drives. The devices are linked together on a bus using a DeviceNet cable with 4 wires: Vcc, CAN-high, CAN-low and Ground. The cable is shielded and has a trunk length of up to 50 meters.

A basic network is comprised of a master such as a PC or PLC and a set of simple devices such as on/off switches or motor starters.

Object-oriented techniques (OOT) are used where objects are defined in class, instances and attributes. This allows changes to be made more easily. Four basic objects are used. An identity object provides information about the identity of a device. A connection object contains the number of logical I/O ports of the device.

A DeviceNet object configures the port attributes such as the node address and data rate. It maintains the configuration and status of physical attachments. You can remove devices from the network or insert them into the network without removing network power. You can also add power taps at any point in the network and have redundant power supplies. One goal of DeviceNet was to achieve the same level of interchangeability for 120/220V-ac and 24V-dc discrete devices using digital communications.

DeviceNet provides interchangeability of simple devices while making interconnectivity of more complex devices possible. Instead of a source-destination approach, DeviceNet uses the producer/consumer model, as provided by CAN, where each data packet has an identifier field. The identifier provides for multiple priority levels for arbitration and multiple consumers.

The device with data produces the data on the network with the proper identifier and all devices needing data listen for messages. When devices recognize the appropriate identifier, they consume the data.

The message length for DeviceNet is 0-8 bytes, which is ideal for devices with small amounts of frequently exchanged data. DeviceNet supports strobed, polled, cyclic, change-of-state and application-triggered data movements. Communication can be master/slave, multimaster with different slaves, peer-to-peer, or a combination configuration depending on device and application requirements. Response time for a network of up to 64 nodes and 500 I/O can be less than 5-ms.

Devices can be powered directly off the bus and communicate using the same cable. Nodes can be configured, removed from, or inserted to the network without powering down. Power taps can be added at any point in the network allowing redundant power supplies.

Trunk and drop lines can use thick or thin cable. The network distances depend on data rate and cable size. A range of over 500 feet is possible.

In DeviceNet's communication stack of OSI layers, the CAN protocol defines the physical and data-link layers. DeviceNet adds the remainder of those layers, along with the media and application layers.

A DeviceNet network can have as many as 64 node addresses. Since it uses the CAN data-link layer, DeviceNet is a producer-consumer model, which requires packets on the network to include data-identifier fields. This approach is in contrast with older technologies where messages include a explicit source and destination. The identifiers tell the devices if they should consume the data.

The network assigns transmissions associated with a connection device configuration. The messaging protocol indicates how a device should interpret a message. A message router transfers received messages to target objects. The priority of access depends on the Message Group the device belongs to. Device configuration takes place through the device connections when the device identifies it.

DeviceNet Objects and Devices

The object model provides a template for organizing and implementing the attributes (data) and service (methods or procedure). An identity object's attributes include vendor ID, device type, product code, revision status and serial number.

To facilitate compatibility and interoperability, DeviceNet defines standard device profiles, designed to allow you to interchange vendors and products that conform to the same device-type profile.

DeviceNet provides a communications link for connecting devices such as sensors and motor starters to a network. DeviceNet uses two twisted pairs of wires to provide 24-VDC power and signal. One survey by Venture Development Corporation found that DeviceNet was used by 44% of instrument and control device suppliers.

DeviceNet Future

U.S. device/sensor buses are projected to grow to 2.7 million nodes in 5 years. DeviceNet should account for over one-third of this total.

More than 300 ODVA members direct the future of DeviceNet through special interest group activities. These include alternative media options such as noninsulated flatwire cable, to reduce costs, and gateways and high-speed routers to other networks.

When two or more nodes try to access the network simultaneously, an arbitration scheme resolves the possible conflict with no loss of data or bandwidth. This is much different than Ethernet where collision detection is used where a loss of data and bandwidth can occur since both nodes have to back-down and resend the data. If two nodes transmit simultaneously, the winner is the one with the lower numbered identifier.

Cyclic redundancy checks are used by CAN controllers to detect frame errors. Automatic retries are also used. These methods are usually transparent to the application and serve to block a faulty node from upsetting the network.

CAN

The controller area network (CAN) was developed by Bosch to replace expensive harness wiring. The technology of the CAN bus was devised in the mid-1980s as an in-vehicle network for automobiles. Since the 1980s, the automotive industry has been using electronic control systems for in-vehicle functions and features that improve safety and comfort.

In a system with conventional wiring, data are exchanged using dedicated signal lines. As the complexity and number of devices increased, the use of dedicated signal lines became increasingly difficult and expensive to implement.

CAN is used as the Data Link Layer of DeviceNet. CAN controller chips allow data to be moved on DeviceNet using a data frame. CAN chips are available from Phillips, NEC, Intel, Siemens and National Semiconductor. A 0-8 byte size is used for low-end devices where small amounts of I/O data that must be exchanged regularly. Simple devices can also send diagnostic data.

Bus systems like CAN overcome some of the limitations of conventional wiring. Using the CAN bus, devices are connected to a common serial bus. Any device on a CAN bus network can communicate with any other device using a common pair of wires.

CAN bus features include CAN frames which have identifiers at the beginning of the frames. This identifier is used to identify which device should receive the frame. Along with the identifier, other pieces of information are added to the frame, such as data length, data bytes, error checking, and frame acknowledgment.

One type of CAN network is called a producer/consumer network. This type of network allows any device to produce or send a message to any other device. This can be done without any interaction from a host or master. This type of network decreases the amount of bus traffic.

When a device transmits a message onto the CAN bus network, all other network devices receive that message. Each receiving device conducts a test on the incoming frame identifier to determine if the message is intended for that device.

When transmitting, a device examines the network to ensure that higher-priority identifiers have precedence using a collision-handling system called non-destructive bitwise arbitration.

The CAN bus provides a physical layer that is implemented through CAN controllers and transceivers. It is a multidrop network with no limitation on the number of devices, provided that the electrical characteristics of the signal remains within specifications. The CAN bus cable carries both the power and signal.

The CAN bus defines the hardware requirements and how data frames are transported using a communication medium. A higher-level protocol (HLP) is needed for communications management to set the rules, behavior, and format of the data frames. HLPs for CAN bus technology include DeviceNet, CANopen, Smart Distributed Systems, CAN Application Layer (CAL), and CAN Kingdom.

CANopen

CANopen is a set of specifications with a communication profile as well as device, interface, and application profiles. The communication profile is based on the CAN data-link layer and the high-speed physical layer specified in ISO 11898. CANopen was initially developed in the ESPRIT/ASPIC project by Bosch. In 1994, CANopen was transferred to the CAN in Automation (CIA) international users and manufacturers group based in Germany. CANopen has been used as an open network in production lines, power plants, robot control, medical equipment, military systems, and building automation.

Standardized CAN Networking

In the early years of CAN, each CAN-based embedded network had its own protocol. CANopen deals with the CAN-specific details such as bit timing and implementation. It provides standardized communication objects (COBs) for real-time data, configuration data and network management data.

Besides standardizing communication objects, CANopen specifies the description of application objects. The CANopen device profiles define application objects for the basic interoperability of CANopen devices. To test this interoperability, there is a CANopen conformance test tool. CIA uses this tool for certifying CANopen devices.

In CANopen, the communication objects for decentralized control applications are standardized and described in an object dictionary. The object dictionary allows CANopen to distinguish more than 2,048 objects.

Communication Objects

Process data objects are mapped to a single CAN frame using all 8 bytes of the data. Each object has an identifier and may be transmitted by only one node, but it can receive from more than one.

Service data objects transmit configuration data, which may be longer than 8 bytes. The transport protocol allows the transmission of objects of any size.

Network management objects consist of the boot-up object, network management object, and error-control object. The boot-up object has the same identifier as the error control object and is transmitted after initialization before the node is set into a preoperational state. The network management object is the highest priority object in a CANopen network and is transmitted by the network management master node.

Sync, emergency indication and time-stamp transmission objects are broadcast periodically by the sync producer. The sync object provides the network clock. Emergency objects are triggered by a device fatal error and are transmitted from an emergency client on the application device.

ControlNet

ControlNet is a real-time, deterministic control-layer network. It can provide high-speed transport of real-time I/O and messaging data. It allows peer-to-peer messaging on a single physical media link and can be used for configuration and programming which can be done off-line with a hand-held device.

ControlNet tries to combine the benefits of several networks onto one link for increased system performance. It is designed as a control network to reliably predict when data will be delivered and ensure that transmission times are constant and unaffected by the devices connected to the network.

As a real-time, control-layer network for the high-speed transport of time-critical I/O and message data, ControlNet provides uploading/downloading of programming and configuration data and peer-to-peer messaging, on a single physical media link.

ControlNet operates at 5-Mbits/second. It allows multiple controllers to control I/O on the same wire. This is an advantage over other networks, which allow only one master controller on the wire. ControlNet also allows multicasting of both inputs and peer-to-peer data, which reduces traffic and improves system performance.

Determinism is the ability to reliably predict when data will be delivered. The deterministic performance of ControlNet can be used for both discrete and process applications. Other features include user selectable I/O and controller interlocking update times to match application requirements.

Network access is controlled by a time-slice technique called Concurrent Time Domain Multiple Access (CTDMA), which regulates a node's chance to transmit in each network interval. The minimum interval time is 2 ms. ControlNet is useful for systems with multiple PC-based controllers.

ControlNet can function in environments where responses occur within milliseconds. Each time a user adds a device to a network, the cycle time slows and performance is affected. ControlNet is based on a producer/consumer communications model and can schedule the delivery of priority messages in spite of the number of nodes.

Producer/consumer Models

Networks can be grouped into one of two categories; source/destination or producer/consumer. In a source/destination network, messages are passed from the source to each destination at a time. With producer/consumer, the producer broadcasts the message to every consumer at once. In the network environment, devices use identifiers which are embedded into each message to determine which messages they should consume. The source/destination model does not impact the baud rate at which data are transmitted—it affects the bandwidth. Since the producer/consumer network transmits a piece of information only once, it uses less bandwidth.

Redundancy and Security

Controller redundancy allows users to specify a backup for any number of nodes or connections. Until the backups are needed, the nodes are in listening mode and can take over for a failed device without powering down the system. ControlNet provides physical level redundancy.

Guaranteed bandwidth means that critical data should not be lost. In ControlNet a user can configure a temperature sensor to update every few milliseconds to control proportional-integral-derivative loops.

Most source/destination networks can only send I/O updates as fast as polling the network allows and all nodes are updated at the same rate. The data may come too late for some uses and too often for others.

Future Directions

Future advancements for ControlNet include conformance testing, reductions in the size and cost of ControlNet chips and real-time I/O (implicit messaging) to ControlNet services over Ethernet.

Open systems like ControlNet need products that are tested to ensure they are compliant with the network's specifications. Any network like Profibus, ControlNet, or Foundation Fieldbus, can have interoperability problems and there could be difficulty integrating products from multiple vendors.

The ControlNet chip is called the SMAC4. The size of this 5-volt chip should shrink to 0.6-micron from the 1-micron size and costs should drop as a result.

ControlNet and Ethernet

Ethernet by itself does not provide interoperability and/or interchangeability among competing devices because a standard application layer, common object model and universal device profiles do not exist.

TCP/IP is used to apply ControlNet messaging services over Ethernet. This method treats the Ethernet network as a limited-function ControlNet network. The technique allows a ControlNet node to encapsulate a ControlNet message as the data portion in an Ethernet message.

The node can then send the message using TCP/IP with the ControlNet message inside. The message is sent to an Ethernet communication chip on the data-link layer instead of a ControlNet chip. Since the control and information protocol is used as the application layer, ControlNet's object model and device profiles can be used.

TCP/IP is basically point-to-point so ControlNet services over Ethernet are limited to explicit messaging only. The data field of each message carries both the network's protocol information and explicit instructions for service. Nodes must interpret each message, execute the requested task, and generate responses. ControlNet can simultaneously transfer explicit messaging and real-time I/O, with UDP/IP. Unlike TCP/IP, UDP/IP is not limited to point-to-point communications and can be used to multicast messages.

ControlNet's native data-link layer uses concurrent time domain multiple access to provide guaranteed access for scheduled I/O. Ethernet uses carrier-service multiple access/collision detection.

Seriplex

This network has been used to control the signaling of sensors by General Motors. GM also uses hundreds of Seriplex I/O points in monitoring and material call systems.

The Ford Assembly Plant in Toronto uses Seriplex to link its lighting control system to meet the energy reduction goals of the Ford 2000 environmental-impact program. The Seriplex system links 200 panel-boards to a central PC which controls the entire lighting system. Replacing the point-to-point wiring, Seriplex made it possible to retrofit this 1952-era plant with PC lighting control with minimal wiring and installation time.

Seriplex was developed by Automated Process Control (APC), in 1987. Square D bought APC in 1995 and created the nonprofit Seriplex Technology Organization. This device-level control network has been used in North America, Asia, and Australia with more than 3 million I/O points installed at more than 3,000 locations. Besides Square D, Seriplex manufacturers include Turck, Pepperl+Fuch, ABB Flexible Automation, Banner Engineering, Delta Automation, GE Fanuc Automation, MAC Valves, and Wonderware.

Seriplex technology was designed mainly for sensors, actuators, and simple on/off devices. It is deterministic, with speeds in the low milliseconds.

A single network can accommodate several thousand I/O devices and extend more than 5,000 feet. The network allows for the control and monitoring of analog and digital devices.

Seriplex chips are embedded in sensor and actuator devices. This provides the connection to the bus. The bus is a serial, multiplexed, intelligent, distributed I/O system with master/slave and peer-to-peer I/O control and logic. Seriplex technology can replace large bundles of point-to-point wiring with a single thin cable.

Seriplex can reduce downtime and improve safety. It allows operators in a central control station to quickly locate the source of breakdowns.

Seriplex and Ethernet

Seriplex may be able to be paired with Ethernet. This would replace multiple levels of proprietary and often incompatible networks with a system of open technology. Several trends could make this possible.

One is the emergence of Ethernet and TCP/IP as de facto standards. Most intelligent devices are compatible with Ethernet. TCP/IP is the technology that makes it possible for computers to view data over the Internet.

As Ethernet moves from a shared to a switched design, it can participate in time-critical functions for automation and control. The

MODBUS protocol could be used for exchanging data.

Ethernet interfaces smoothly with device networks such as Seriplex and AS-i (Actuator Sensor Interface), which is used in Europe. Both of these are deterministic, which allows them to maintain control sequencing and accuracy.

Seriplex networks can accommodate up to 7,680 discrete or 480 analog I/O points, with up to 300 devices connected to one network cable of up to 5,000 feet. AS-i can accommodate 31 nodes and up to 248 I/O on a 330-foot network or 660 feet with a repeater.

Filters can be used to monitor which I/O points report automatically, such as when an emergency stop is needed. The filter can also be used to capture and record data.

Advancements in fieldbus technology are having a major impact on the growth of faster, smarter, and more efficient industrial networks. Higher-level protocols such as Profibus offer the high performance levels required in automation applications. The lowest level of automation is the actuator and sensor level where the AS-Interface (AS-i) is used (Figure 8-1).

AS-i

This standard was invented in 1990 as a cost-efficient alternative to the conventional hard wiring of binary field devices. As-i has been used in hundreds of thousands of installations. Originally, AS-i focused on the connection of low-level binary devices with a single, easy-to-mount, two-conductor cable that carries power and data.

AS-i is now an international standard under EN 50295. It is a vendor-independent bus network that is open to any manufacturer. AS International is the sponsor organization of AS-i. It was formed in 1991 by a consortium of 11 manufacturers who developed the international standard. Today, the association has over 100 member companies that supply more than 600 products and services. AS International is also responsible for the conformance testing and certification of devices.

Connection to Higher Levels

AS-i provides a low-cost solution for connecting binary products, such as sensors, actuators, valves, and relays, with higher-level

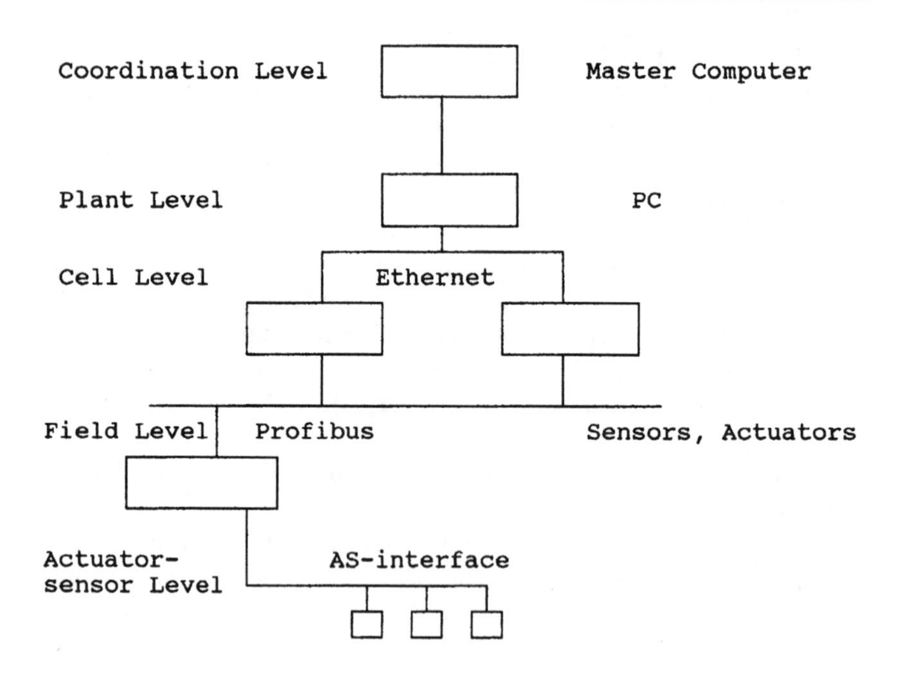

Figure 8-1. Network hierarchy.

control devices including PCs and distributed control systems (DCSs) (Figure 8-2). It is used at the level directly below fieldbus and device-level network systems. It transmits control and device data, configures the system architecture, powers the devices, and monitors the network.

AS-i can be used for upgrading HVAC systems. Parallel control wiring can be replaced with a simple AS-i cable (Figure 8-3). Most conventional wiring system can be replaced and users report wiring savings of up to 40%.

Conformance to the AS-i standard assures interoperability. Each slave that fulfills the standard's requirements can work together with any other slave in the network.

The network uses a common cabling system with all peripheral devices connected to the AS-i cable. The AS-i protocol is embedded in the slave chip which connects the sensors and actuators to the cable.

AS-i uses one master per network that calls up all the connected slaves with a cyclic polling that uses their address. The AS-i data packets are short with simple structures and have a fixed length. Four usable data bits are exchanged between a master and every in-

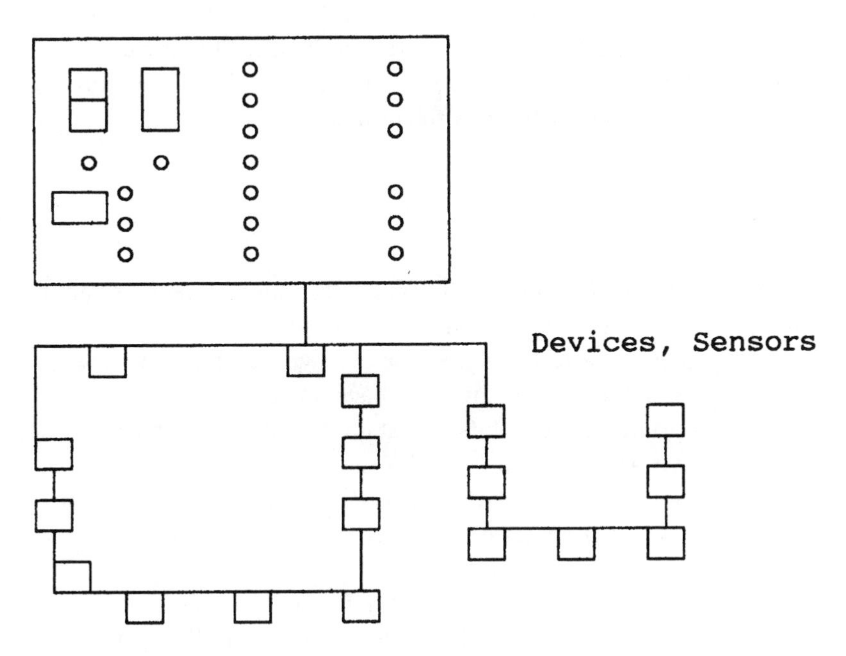

Figure 8-2. Cabling with AS-i serial interface.

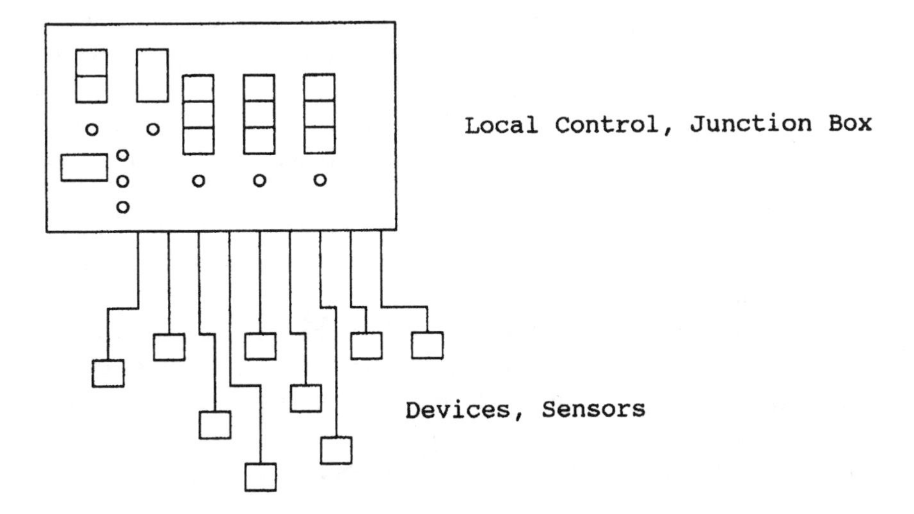

Figure 8-3. Parallel wiring cabling.

dividual slave during one cycle. Updated data of the running process can be found in the master as well as in the slave after one cycle. In a fully expanded system with 31 slaves, the cycle time is about 5-ms.

The cyclic polling method is deterministic. This means that after 5 ms, updated sensor data are available to the control unit and is then transmitted from the control unit to the actuators.

System Operation

The AS-i slave integrated circuit (IC) allows sensors, actuators and modules to link up as slaves to the AS-i bus. The sensor or actuator gets its power from the line, recognizes the data the master sends out and then sends the data back to the master.

During each cycle, 4 bits of data are transferred from the master to each slave and then back to the master. The ports can be configured as inputs, outputs, or bidirectional. Each time a chip is installed in a slave, the I/O configuration sets the port's configuration, which is a 4-bit, permanently stored value.

AS-i can be scaled beyond the 4 bits of data per cycle to slaves with more than 4 bits of information. To accomplish this, the data transmission is spread out over several cycles. AS-i can also be used for analog sensors and actuators.

The As-i chip may be built into sensors or actuators to make an integrated device, or it can be built into a module where sensors, actuators, and switches are connected.

The AS-i master is a central device that organizes the exchange of data in the network and connects it to a higher level control unit. Three types of masters may be used. Pure masters allow a direct link to a PC. Couplers allow the connection to a field bus. Stand-alone masters include some control functions that let it work without a higher control.

The AS-i master is self-configuring and is configured when the addresses and the individual I/O configurations and ID codes of the connected slaves are written into the master. The master automatically performs all noncyclic functions that are needed for parameters, automatic addressing and network supervision.

Gateway Access

Masters that behave as a slave in a field-bus network act as couplers or gateways. These AS-i gateways provide access to sensor and actuator data from all layers of the automation architecture. AS-i gateways are available for Profibus, Interbus, DeviceNet, CAN, SDS, MODBUS, MODBUS+ and RS-232/-422/-485. Stand-alone masters act as controllers for simple tasks.

An As-i power supply unit serves as the power supply for the slaves. It forms the voltage pulses for data transmission and can be connected to the network at any point. It feeds a symmetrical supply voltage of 30 VDC into the cable. Power supplies are available with ratings of 2.2 to 8-A.

AS-i Cable

The standard AS-i cable is 2-core, 2 by 1.5-mm, rubber insulated, with a special profile that prevents reversal of the poles when connecting devices. The electrical connection is made by penetration contacts, which pierce the rubber insulation to make a connection to the two cable wires. This connection method eliminates stripping and terminating.

If a device needs to be moved to another location on the cable, it is easily done. The cable is self-healing and holes made in the cable during installation close when the connection is removed.

Cables are available for applications that require data and power separately. A black cable is used for supply voltages up to 30-V and a red version for supply voltages up to 230-VAC. Table 8-1 lists the AS-i cable characteristics.

Table 8-1. AS-i Cable Characteristics

Trapezoidal mechanically coded cable
Reverse voltage protection
Oil-resistant versions
UL recommended
Data and power, yellow version - 30 VDC
Power - 24 VDC (black) to 230 (red) VAC
Piercing connectors

Network Structure

The AS-i network uses a tree structure with a maximum of 31 slaves, to which 124 inputs and outputs can be connected. In most applications, the master is part of the first control level, but the network can also be subordinate to a fieldbus system.

The master becomes a slave in the fieldbus network. It controls and polls the network, sending and receiving data from each connected device. It senses and registers any connected device without configuration or application-specific software. There are couplers for different fieldbuses, including Profibus, Interbus, CAN, DeviceNet, and SDS.

Over a million AS-i field devices have been installed. There are slave modules with integrated pneumatic valves, earth fault detectors and slaves to handle analog signals. Repeaters and extenders can be used to increase the 100-m network length.

The number of slaves that can be connected to a network has been increased from 31 to 62. This is done with A/B addressing, which uses a spare bit in the protocol to switch between devices in succeeding bus cycles. This increases the maximum network capacity to 248 inputs and 186 outputs but doubles the cycle time to 10-ms.

A spare bit in the status register is used to mark peripheral faults. The host can use a fast fault indicator, making it easier to detect faults in a network.

HART

HART is a de facto standard in the automation industry. HART offers ease of use, compatibility with existing systems, and cost-saving benefits. HART retains the compatibility and familiarity of 4-20-mA systems, while providing remote two-way digital communication with smart field equipment. HART is an important communication technology used in smart instruments.

Smart instrumentation is important in distributed intelligence and control. The ARC Advisory Group has predicted that HART will dominate smart-instrument communication. There is a major trend from traditional analog to smart instrumentation.

The HART Communication Foundation (HCF) owns the technology, which was developed by Rosemount. HCF is a nonprofit organization with worldwide members. Sales figures indicate that HART is the leader in South America, very popular in Asia and becoming the industry standard in Europe. The HART protocol is expected to have an installed base of 1.5 million in several years.

HART has the capability of allowing multiple measurements and calculations in one device including diagnostics. It is a straightforward and simple technology that provides low-cost digital communications.

In most applications, HART digital information is communicated simultaneously with the traditional 4-20-mA analog signal. The variable is transmitted continuously via the analog signal, while additional measurements, calculated values, device configuration, diagnostics, and device status information are sent by HART digital communication on the same wires.

Another capability is that several HART-compatible field devices can be connected to a single pair of wires in a multidrop network. In this system, digital communication is used to access all of the device information.

In the future, there will be more network-related products, multiplexers, and multidrop connections for on-line interaction with HART devices. The digital communication capability allows device information to be linked remotely to other digital networks. HCF has developed an object linking capability for process control (OPC) servers that allows cyclic access of HART device information by higher-level systems and client applications.

HART-compatible multiplexers and I/O subsystems make it easy to connect HART devices with central control systems, maintenance stations, and control and data acquisition networks through the OPC Server. The server provides cyclic data access to process variables and device diagnostic information. Other digital networks for industrial automation include Foundation Fieldbus and Profibus.

Foundation Fieldbus

The Fieldbus Foundation has developed Foundation Fieldbus. The goal was to develop a digital fieldbus that meets the control requirements of process applications. It is based on ISA and IEC standards, as well as standards that cover Profibus, the French FIP, (factory-instrumentation protocol), and HART (highway addressable remote transducer), a common open protocol for industrial automation.

Fieldbus devices and host control systems provide an alternative to analog control. Fieldbus is an all-digital, serial, two-way communications system to interconnect sensors, actuators, and controllers. It serves as a local-area network for instruments used in control and automation applications. Fieldbus provides reduced infrastructure requirements especially in the area of field cabling. Each fieldbus segment is capable of supporting multiple devices. The expensive cable run between the field-device junction box and the control system will be much smaller.

Since Fieldbus can drop spurs along its length, this reduces the number of home run cables required. A digital signal can also be superimposed over an analog signal which carries the control variable measurement.

Savings can also be achieved with fieldbus multivariable transmitters. These allow a single device to transmit more than one variable. The signal to be multiplexed can be temperature or pressure measurements.

Fieldbus and digital signals also offer increased reliability. This increased reliability comes from the use of the diagnostic capabilities of the system. The network transmits this information to the network, where it can be used to monitor the health of the system.

According to some studies, when a problem occurs, 2/3 of the repair time is spent in diagnosis and only 1/3 goes to the actual repair or replacement. Reduction of the time needed to diagnose the problem can have a significant impact on maintenance costs.

Foundation Fieldbus technology comprises the physical layer, the communication stack, and the user application. These components fit the OSI communication model. The physical layer allows low-speed devices to run on twisted-pair cable.

The link active scheduler (LAS) is in charge of access to the bus. A link master is a device that can be an LAS. The LAS capability can be transferred to secondary link masters, providing redundancy for the system.

Fieldbus defines an eighth layer, the user layer. The user layer defines blocks that represent the functions and data in a device.

You can implement function blocks on the fieldbus for PID and PD functions which could each reside in separate devices, such as a transmitter, a loop controller, and a valve.

In these systems, Fieldbus provides a type of distributed control, in which the control function exists in the field rather than in larger central controllers. A control-system host with a Windows NT-based human machine interface can be used to create an interface for configuring device parameters.

The digital circuitry of Fieldbus transmitters improves the accuracy of sensing and transmitting the control variable. Just as Windows has grown from its initial versions to its present architecture, Fieldbus should continue to grow and expand. One of the features added to Fieldbus is the ability of devices to monitor revisions of the device description to ensure that it is compatible with the one in the host system and others on the network.

Profibus

The compatible protocol variations in Profibus include Profibus-FMS, Profibus-DP, and Profibus-PA. Profibus-FMS handles high-level, non-real-time communications among devices. Profibus-PA targets process-control applications.

Profibus-DP is a low-level network that targets high-performance I/O scanning. Typical DP-slaves are discrete inputs or outputs for 24V dc or 230V ac analog inputs, and analog outputs.

Profibus-DP provides high-speed data transfer at the sensor and actuator levels. Controllers, such as PLC devices, Master devices, or active stations can control the bus and transfer messages without a remote request. Slave devices are simple peripherals, such as sensors and actuators.

The Profibus standard defines two physical layers with medium-access protocols for different transmission techniques. Besides the physical layer cable a second physical medium is fiber-optic, which greatly extends the bus length at high transmission speeds. The Profibus data-link layer is designated as the fieldbus data link.

CAB

In 1985, Public Works Canada conducted a study for a more cost-effective means of operating and managing the control systems used in public buildings. This development focused on an operator workstation that would simplify building operations and reduce operator training. A new data communications protocol was required to link this workstation to the building automation systems in use. The Canadian Automated Building (CAB) Protocol was supported by Public Works and Government Services Canada (PWGSC) and compatibility with this protocol is now required for all new building control systems procured by PWGSC.

CAB uses a Local Area Network between gateways and workstations. It defines a standard interface console that connects to the LAN and communicates to nodes via defined messages. CAB provides user interface capabilities across multiple-building systems.

CAB is essentially a high-speed network protocol and best used when communicating between devices with good processing capabilities. It is rarely used for field-device communication.

MODBUS

MODBUS is an Instrument Society of America (ISA) protocol that has been developed for factory automation. It allows communication between different manufacturers' programmable logic controllers (PLCs) and intelligent field devices, such as transmitters and actuators.

Devices that use MODBUS typically require healthy processing capabilities, and are usually more expensive than the control devices used in a typical commercial or institutional building for environmental control.

From proprietary to public standard, the proliferation of industrial networks offers user benefits, but unfortunately many of the benefits are spread across different networks.

Plant-wide network architectures are evolving into three layers. ControlNet fits into the middle automation and control layer for real-time I/O control, interlocking, and messaging. The upper information layer for plant-wide data collection and program maintenance can be done with Ethernet. The device layer for the integration of individual devices is handled with the device buses like DeviceNet and Foundation Fieldbus.

References

"A Host of Options at the Device Level," *Worldbus Journal,* (Supplement to *InTech*), Vol. 46 No. 10, October 1999, pp. 38-42.

Amborn, Randy, "Standard for Open Protocol has Several Advantages," *Contractor*, Vol. 41 No. 6, June 1994, p. 24.

Bryant, Michael, "A Little Yellow Cable Makes Big Connections," *Worldbus Journal*, (Supplement to *InTech*), Vol. 46 No. 10, October 1999, pp. 28-31.

Devoney, Chris, "The Unwired Nation," *Windows Sources*, Vol. 2 No. 4, April 1994, pp. 108-138.

Gahran, Amy, "BACnet on Duty: New BAS Frontiers for End Users," *Energy User News*, Vol. 23 No 3, March 2000, pp. 16-21.

Helson, Ron, "Get the Best of Both Worlds with HART, *Worldbus Journal*, (Supplement to *InTech*), Vol. 46 No. 10, October 1999, pp. 20-21.

Hull, Gerry G., "BACnet: Miracle or Mirage?, *Heating, Piping, Air Conditioning*, Vol. 69, September 1, 1997, pp. 38.

Moss, William H., "ControlNet Speeds Up Data Transport," *Worldbus Journal*, (Supplement to *InTech*), Vol. 46 No. 10, October 1999, pp. 36-37.

Newell, Edwin R., "Seriplex Signals in New Era," *Worldbus Journal*, (Supplement to *InTech*), Vol. 46 No. 10, October 1999, pp. 33-34.

"Single Control System in Buildings Possible with New ASHRAE Standard," *Air Conditioning, Heating and Refrigeration News*, Vol. 196 No. 5, October 2, 1995, p. 14.

Verhappen, Ian, "Dawning of the Digital Age for Process Control," *Worldbus Journal*, (Supplement to *InTech*), Vol. 46 No. 10, October 1999, pp. 8-9.

Watkins-Miller, Elaine, "To Interoperability... and Beyond," *Buildings*, Vol. 91 No. 3, March 1997, p. 24.

Internet: www.elibrary.com/s/edumark, GSA Implement BACnet in Golden Gate Project, (General Services Administration; Building Automation and Control), *Heating, Piping, Air Conditioning*, Vol. 69, March 1, 1997, p. 8.

Internet: www.buildingsmag.com/magazine/aug_1999/lonwork.asp, Patterson, Maureen, "The Language of Efficiency," August 1999, pp. 1-5.

Internet: www.trane.com/commercial/equipment/forward.asp

Shinn, Joe, "Moving Forward with BACnet," March 8, 2000, pp. 1-11.

Chapter 9

Building Automation, Wireless Technology, And the Internet

The business of buildings requires the vision to anticipate future needs and deliver them. Conveniences may include plug-and-play wiring, lighting, and air distribution.

The tenants may want to call the building on a Sunday night and program it to be running when they arrive for some needed after-hours use. They may also want an integrated security system that makes them feel safe without invading their privacy.

These things are possible now and will become mandatory in the future. This increased functionality does not have to cost more. Many future products may be made for a building communications technology like BACnet or LonWorks. This type of bus technology allows products of various building systems to communicate.

It can get you the choice of products within a building and competitive bidding within an organization. This choice, or flexibility, equates to the ability to specify vendors, products and solutions

BACnet Projects

Great variety is evident in the BACnet projects that have been done or are under way. In one school district's main energy management system, a BACnet gateway allows the district's energy management system to transmit a setpoint to one or more of the rooftop unit controllers.

In another project, BACnet is the key to communication between an ice rink's BAS and the controls for the chillers that make the ice in the rink.

The city of Memphis, Tennessee has installed new control systems as well as upgrading and expanding its Fairgrounds Complex. The retrofitted controls are BACnet-compatible, and will be networked into a centrally monitored system via the city's existing municipal stages. The flexibility offered by BACnet appeals to city managers.

BACnet is being used in thousands of building automation systems operating around the world but that is only a small portion of all controls projects being done. One of these projects is a centralized network of BACnet-based building automation systems in Tucson, Arizona for about 25 municipal facilities. The new system replaces a group of 5- to 10-year-old direct digital control systems. The city was concerned about their existing control systems, and were expecting to see major system failures if they did nothing.

The project uses BACnet at the head end and the main distributed processors, connecting all of the buildings into a single centralized system. This will allow flexibility to allow true competitive bidding for control system additions in the future.

As more control vendors offer BACnet products, the dream of interoperability through BACnet will be put to the test. The projects that are being planned and executed will be watched closely. One way or another, the era of proprietary BAS protocols is coming to a close.

The U.S. General Services Administration (GSA) is using BACnet controls from Alerton Technologies and the Trane Company at its 450 Golden Gate Building, located in San Francisco. This is the first large-scale implementation of the BACnet open communications protocol. The building is the first large demonstration of BACnet among multiple vendors.

The building houses several government departments and agencies, including GSA and federal courts. It has 22 floors with 1.4 million square feet and is one of the largest buildings in the San Francisco skyline. Each floor is bigger than a football field. The building has seen extensive renovations, including asbestos removal and mechanical system upgrades, while maintaining its pneumatic control system.

The new building-wide energy management control system cost $3.5 million and should save over $500,000 in energy costs per year.

The Alerton system is based on the company's BACnet system called BACtalk[R], and the Trane system is based on the Trane Tracer summit system. The building owners workstation network uses a BACtalk for Windows NT server with an Ethernet 10Base2 LAN to connect the system workstations and network controllers. A network hub is used on each floor with remote terminals connecting to the network via the Ethernet LAN.

Alerton retrofitted over 800 dual-duct and 60 single-duct VAV terminal units with BACtalk controllers. Each controller is programmable and communicates on the BACnet MS/TP LAN. The pneumatic operators have been replaced with electronic actuators and the pneumatic thermostats replaced with intelligent digital-display wall sensors. The eight main dual-duct air-handling units have been retrofitted with programmable controllers.

Trane replaced the pneumatic controllers at each terminal unit with DDC VAV controllers on three GSA floors. Almost 200 dual-duct and 16 single-duct VAV terminal units have been retrofitted. Trane replaced pneumatic thermostats with electronic DDC zone sensors and provided building control units, which are BACnet-compatible. The building level controllers will coordinate the VAV operation and communicate with the Alerton system via BACnet, allowing the two systems to interoperate.

The Windows NT workstations use a graphics package so the operator can view and change the system information. Plans to extend the BACnet system include lighting control and central plant equipment retrofits. Thomas Lighting Controls and Synergistic Control Systems joined Alerton and Trane as manufacturing partners in this project.

Some HVAC systems like the one in the Knickerbocker Hotel in Chicago have reduced operating costs by automating the building's mechanical systems with a new LonWorks-based building control system.

Interoperable LonWorks-based products included ABB adjustable frequency drives, York direct-fire absorption chillers and RCS temperature sensors. Sharing a single, common twisted-pair (78-kbits/s) wire backbone among elements, the system devices work as intelligent zone controls.

Instead of individual wires, the functions are handled by the transmission of messages over a network. The plug-and-play device characteristics allow the connection of the HVAC systems to be done easily on site.

The automation involved multi-zone air-handling devices, refrigeration equipment, and the chilled and hot water systems. The volume of air going into the lobby area is controlled by variable frequency drives. The control of dampers and zones is done with LonWorks-based intelligent actuators. In common areas such as the ballroom, the lobby, and the restaurant, intelligent space sensors are used to transmit network messages back to the actuators that control the zone dampers. In the air handling equipment used, the separation of the hot and cold deck dampers caused a synchronization problem which was solved by the timing facilities of the neuron chips in the actuators. This was adjusted so that both dampers would always operate in the correct positions. The operation of the heating and cooling valves are also under LonWorks control.

The system operates in a Windows-based environment and allows the monitoring of control system data, reset fan temperatures, turning equipment on and off and adjusted operating parameters. Dedicated channels on the CATV backbone are used for guest services such as monitoring the mini-bar, controlling the heating, and providing real-time security with a door locking system.

Products with open protocols embedded in them are more flexible. For example, the retrofitting of a lighting system with occupancy sensors in the building can be done to provide more natural light and decrease the artificial light usage.

Lon-enabled products provide multiple sources of application-specific controllers and programmable devices. These basic elements provide answers to integration requirements.

Products that are cheaper and faster to manufacture are the trend for the future. The benefits of LonMark products include the choice of vendors, easy integration, additions and changes and reduced installation costs.

There has been a tremendous shift in the implementation and utilization of control systems. This shift has been from discrete sensors and analog devices to more intelligent, processor-based networks of sensors and devices that return both data values and diagnostic information.

These intelligent automation solutions are based on device bus networks. Open PC technology such as Ethernet and OLE for process control (OPC) is coupled with device bus networks. This allows desktop, industrial, and laptop computers to be powerful connection tools.

The future requirements for control systems demand that an ever increasing amount of information be provided. The trend is towards open architecture-based technology that replaces proprietary equipment. This force has sparked a era of innovation that has led to the leveraging of existing PC technology.

In today's energy management climate, increasing costs and the need for improved efficiency of building systems forces owners to take a hard look at their building automation systems. Many industries must comply with new environmental and safety requirements.

In such a business environment, less of the required data are control specific. More information of other types is needed to maintain, diagnose and modify the control system.

Standard digital networks have additional benefits. Reduced wiring lowers installation and maintenance costs. Open systems let users select the right equipment for the job, regardless of the system manufacturer. The networks also have intelligent instrumentation, in which better performing devices provide functions such as advanced diagnostics.

The networks offer distributed control with intelligent devices that provide the flexibility to apply control centrally or at local processing points for improved performance and reliability.

The available range from low-level systems that gather data from simple discrete sensors to more sophisticated networks in which intelligent field devices implement distributed control.

Web Technologies

Improved energy management means a tighter integration of the entire facility. Information integration at this level requires the ability to link systems of different types, providing information in various formats throughout the facility and often with geographically dispersed plants and offices. Until recently, this has been a difficult and costly problem, but the increasing use of technologies that have matured on the Internet provides some cost-effective solutions. The Internet has successfully unified a variety of computers, including mainframes, midranges, workstations, and PCs. In many businesses, the move toward electronic commerce means standardizing around Internet technologies, including Ethernet, TCP/IP, and the Web. These technologies are used to track not only their facility but also the status of suppliers including energy suppliers.

A tight integration in some areas means that the state-of-the-art is advancing to the point where this level of integration will be a commodity, more like a standard software package than the research project it seems like today. Improved energy management presents many of the same challenges found elsewhere in the enterprise. Communication and networking standards abound. Programmable logic controllers (PLCs), temperature controllers and PCs all have well-defined but sometimes overlapping roles. The longevity of the equipment means that controllers and software do not always reflect the latest technologies, especially in the area of communication capability.

Often, replacing expensive equipment just to improve the controller capability is not practical. So, while PCs are recycled quite rapidly, equipment controllers, or at least the underlying designs of the controller, may be 5 or more years old which is aged in computing years. Enhanced communications capability might justify adding features to existing equipment, such as tracking energy or fuel use.

Using Internet technologies on a local network to create a local Intranet allows you to integrate many of these systems without a significant impact on operation (Figure 9-1). Networking allows the display of new information, as well as the addition of new control features. The Intranet servers are Web servers with off-the-shelf Web browsers, such as Netscape's Navigator or Microsoft's Internet Explorer, combined with new Java-based technology to provide better access.

The use of Web techniques is fairly straightforward. Information is provided to clients by a Web server. On the Internet, a server may receive many millions of requests for information every day and Web servers are typically dedicated and high-powered computers. For a facility Intranet, the number of requests for information is relatively small, so a Web server can run on a standard off-the-shelf PC system equipped with a network interface. These systems can run the user interfaces. They can also function as real-time controllers.

Web enabling equipment allows anyone with a Web browser and the appropriate permissions can access the information. This makes access to control information uncommonly simple. Once the client computer is set up and networked, you just need the familiar Internet address, or URL. This address might take the same basic form as an external URL, such as www.ibm.com. Internally, an address such as heatingSystem/status.html might provide access to the status of a local system.

In a typical Internet Web site, most of the information is static since it does not change until someone physically changes it. Dynamic applications on the Web need to advance this concept by displaying rapidly changing information in near real time. To create, transport, and view

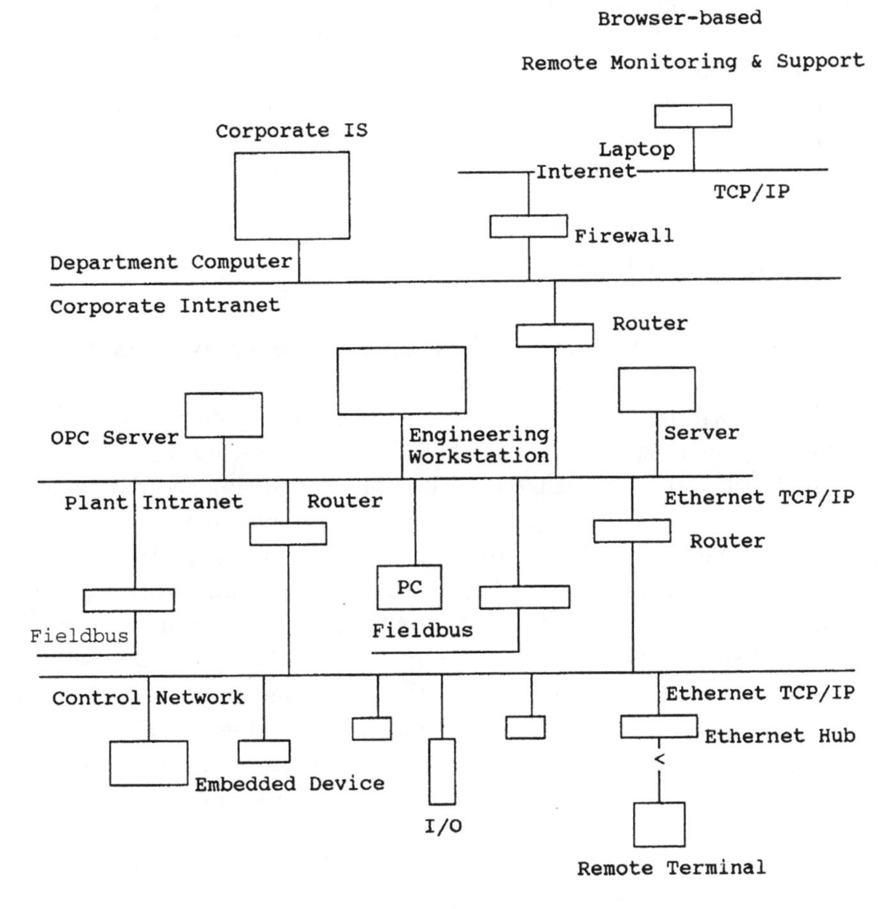

Figure 9-1. Control, plant and corporate networks.

this information, the Java language can be used.

The Java applets are transported across the network and are executed on the user's computer inside the Web browser. The user interface looks familiar to anyone accustomed to Web Surfing.

For the applet to be effective requires that there be communication with the real device for the system to either monitor or control the equipment. The nature of this communication depends on the control function of the equipment.

Since Java and Intranets mainly support only open protocols, this interface is usually widely available throughout the network. One way to create this interface is using the network socket interface provided with most TCP/IP implementations. This provides the data stream from the client to the server.

Sockets are used to talk to Ethernet-enabled PLCs such as Modicon's MODBUS TCP which provides a widely supported socket interface to Modicon and other PLCs.

Distributed Objects

One flexible technique used to communicate across a network is the distributed object. This is a higher-level protocol that is more standardized. Distributed objects allow software running on multiple machines to interact almost as if it were on a single machine.

Distributed object technology allows the processing and data to be distributed to various devices where it may be most needed or most effectively handled, including distribution to the Web applications.

Distributed object standards include the common object request broker architecture (CORBA), guided by the Object Management Group and the distributed component object model (DCOM) which supersedes another earlier Microsoft standard (COM) component object model.

CORBA and DCOM provide similar capabilities for large-scale and facility integration (See Table 9-1). For the users of Internet technologies, the object request broker that enables CORBA, has the advantage of being available on almost every computer platform and operating system. This includes the embedded real-time operating systems used to control machines. DCOM is available only on Microsoft Windows products.

Java provides the networking support for socket programming. A distributed object technique called remote method invocation allows clients using browsers to easily interface to servers supporting these standards (Table 9-2).

Some industries have their own protocols, such as the semiconductor equipment communication protocol, SECS/GEM, or the pharmaceutical equipment communication protocol, LECIS.

Most embedded and real-time computer platforms support Java in the form of Java Virtual Machine (JVM). This is an application or library that converts the portable Java byte code into platform-specific code as it is executed.

Table 9-1. Web Intranet Levels with CORBA

Web browsers with dynamic Web pages
Intranet
Dynamic Web pages and user interfaces - Automation and control
Web server - CORBA
Monitoring and Control
Hardware interface

Table 9-2. Web Intranet Levels with CORBA and JAVA

Web browsers with dynamic Web pages
Intranet
Dynamic Web pages and user interfaces - Automation and control
Web server - Java JVM - CORBA
Monitoring and control - C, C++, Java
Hardware interface

Multiple protocols can run on one controller. These Web-enabled components can run on local Intranets using a choice of open communication protocols.

Intranet networking can take three basic paths:

- Ethernet network enabling,
- Web enabling, and
- Server-side logic.

Ethernet and Web Enabling

If a device has RS-232 communication, a terminal server can be used to convert the RS-232 communication to Ethernet. Terminal servers are available from companies such as Lantronix, Kanematsu and Sierra Monitor.

These systems can also provide some protocol-conversion capabilities, such as from MODBUS RTU to MODBUS TCP. Some terminal servers are also Web enabled. They allow user-supplied Web pages and applets to be downloaded to the terminal server and accessed from there. These applets can communicate with the terminal server and then to the equipment over the Ethernet connection, for monitoring, control and other functions.

One use for these applets is to allow tasks such as configuration, troubleshooting, and maintenance. The manufacturer provides these resident applets. To configure the system, you point a Web browser at the Web server in the equipment or terminal server and configure the system from the pages shown. Troubleshooting can be provided as lists of alarms or fault indications with suggested corrective actions.

The maintenance and troubleshooting can even be handled remotely over the Internet. The use of the remote user-interface can provide information on all aspects of the machine or equipment. The hardware interface provides monitoring of the system and could also control the equipment. Capabilities well beyond the scope of the original controller can be incorporated simply.

There are many PLCs that now include some capability for embedding Web pages and applets within the controller itself or within another module on the controller such as an Ethernet communications module. Future advances in PLCs will enable support for real-time, server-side Java.

This would allow the PLC to provide monitoring and real-time control. Web-based interfaces and integration could be provided directly from the PLC, which would program such logic in Java on both the client and the server.

Using the technologies of the Internet on a local Intranet can provide a cost-effective method of improving integration (Table 9-3). Many new controllers may include embedded Web pages for troubleshooting and maintenance.

Merging Information and Control

In today's deregulated environment, improving plant and facility competitiveness is a driving force behind most HVAC control system replacements. New control systems not only control plant operations, but also improve the access to diagnostic information that can help improve efficiency, reliability and cost effectiveness. This is a change

Table 9-3. Information System Levels

Information from Web Servers and Smart Equipment

Clients with Web Browsers	Servers throughout the facility provide pages containing information from all equipment and database links	
Open protocols		
Facility devices and equipment		
Existing equipment with Web server	Ethernet with terminal server	Smart Ethernet devices
BacNet, DeviceNet, Fieldbus, Lonworks		
PLCs, smart controllers		

from the past, when control systems were typically replaced only after they became so old that replacements parts were either unavailable or too difficult and costly to obtain.

New information technology plays such a key role in nearly all new plant control systems that they are sometimes called information and control systems. These systems integrate the entire plant instrumentation and controls, plant computer, and information systems, along with communications and diagnostic capabilities, into an enterprise management system.

This enterprise management is a plant infrastructure that merges information technology and instrumentation and controls. It acts as a unified information and control system with a shared information storehouse. It focuses on the advantages of information and control technologies to reduce plant costs.

Enterprise management includes operating the plant and tracking operating costs. It also involves the use of modeling that tests different scenarios to determine the impact of control and maintenance actions on operating costs. The traditional boundaries between control systems and information systems are fading.

Building Automation and the Internet

Building automation systems are also using the Internet. Systems that work with the Internet take building automation systems (BAS) online. An interoperable BAS system can send out signals through the Internet to a PC that the HVAC equipment is not working properly. Problems can then be acknowledged and corrected remotely. Weather information could also be gathered from the Internet and downloaded to a building system, which could then adjust the buildings' systems accordingly.

The Internet connection could also be used to check utility prices across the nation on a regular basis and make adjustments according to a building's needs.

Interoperability provides the ability to combine products from multiple vendors into flexible, functional systems without the need to develop custom hardware, software, or tools.

Wireless Technology

Wireless technology will be everywhere in the future, and is likely to be prevalent in building automation. There are smart vending machines that are equipped with wireless modems that notify the warehouse when stocks are depleted, eliminating the need for costly, on-site time-intensive inventory checks. Portable ticket stations, which are terminals linked to printers via wireless modems, are found in many airports.

The type of wireless technology used depends on the maximum distance between the home base and the remote site. If you are roaming a building, wireless extensions, such as bridges or routers, can link you to a local area network. To cover a wider area, such as a campus or building complex, a series of LANs using wireless extensions can be used to transfer any kind of data. In other cases, infrared or low-power radio equipment linking one PC to a portable computer may be used.

Wireless technologies run the gamut from those based on radio frequencies and cellular technologies to newer technology such as cellular digital packet data (CDPD) which is a digital alternative to analog-based cellular communication technology. Any type of information that can be transmitted wirelessly can be sent over a coaxial cable or a telephone line.

One of the most commonly known wireless technologies depends on radio frequency (RF) signals, which send and receive data using low-power transmitters and receivers. RF signals make up a portion of the electromagnetic spectrum and are used in local-area RF and wide-area RF implementations.

RF transceivers and infrared devices are used to create wireless LANs, which can connect PCs on a peer-to-peer network or connect a node to a LAN. The range for local-area RF equipment starts from 30 feet to about 100 yards and some equipment can reach 600 feet to 1/4 mile (1,200 feet). Most local-area RF equipment operates in the Industrial, Scientific, and Medical (ISM) bands, which covers radio frequencies that do not require a broadcasting license. The equipment must meet certain power and bandwidth restrictions enforced by the FCC (Federal Communication Commission).

Most wireless LAN equipment falls into two categories: low-speed devices, which range from about 115 kilobits per second (Kbps) to 250 Kbps and the high-speed devices that transfer data at 1.25 megabytes per second (10 megabits) or greater, which is about the same speed as Ethernet.

Some wireless connections involve RF transceivers that you attach to the serial port on the PCs. These devices transfer data at the rate of the serial port within about a 10-meter range. These small card-sized transceivers can transfer files between a desktop PC and notebook PC. They form a two-node network in which one machine, typically the portable device, can access resources available to the other unit, such as a printer.

Infrared transceivers also clip on to a serial port and can transfer data between ports. This allows the infrared link on any portable system to exchange files with other computers or to access its resources. There are also versions that attach to a printer, letting you send files to the printer via infrared technology. Complete wireless-LANs are available from companies such as NCR.

The NCR WaveLAN PCMCIA network adapter card fits into the mobile user's notebook, then sends and receives RF signals from a network bridge which is connected to the LAN server. PCMCIA stands for the Personal Computer Memory Card International Association. This industry group determines the standards for all credit card-size peripherals, including modems.

A key issue in wireless LANs is security. Most wireless LANs radiate their information in a nonencrypted format, so sensitive information is vulnerable. Encryption should be widely used in the future.

IrDA Control

The Infrared Data Association (IrDA) Control technology is a command-and-control architecture for communication with wireless peripheral devices. It has an operating range of approximately 7-m. This system is oriented toward control data packets. It is not intended to pass files.

The IrDA Control system is a polled-host topology. The host device polls up to eight peripheral devices in an ordered sequence, providing service requests and handling the peripheral-device responses. The host can be personal computers with peripheral devices such as a mouse and keyboard. After the system boots up, the keyboard and mouse operate with the host PC in the same manner as a wired keyboard and mouse.

Portable devices are using low-cost, high-speed, wireless connectivity to support interoperability. Infrared (IR) technology can be embedded into portable devices such as notebook computers, personal digital assistants (PDAs), wireless phones and digital cameras. It can offer valuable HVAC system maintenance features including walk-up interrogation, file transfer and network access. To ensure platform interoperability, the Infrared Data Association communications standard is optimized for low-cost, short-range, point-to-point communications.

The IrDA standard communications protocol is defined to minimize code size and maximize efficiency. There are several operating speeds and modulation schemes.

IR technology in wireless systems offers some advantages over RF technology. These include lower communications subsystem parts cost, smaller physical space and no Federal Communications Commission (FCC) testing and approval requirements. IR-equipped devices also support more secure data transmissions since they provide shorter coverage ranges and IR light cannot escape areas enclosed by walls.

The IrDA standard covers interoperability by addressing the software layers and physical modulation schemes needed for transmission. The hardware includes an IR controller and a transceiver. The controller has the following functions:

- interface to the system bus,
- converts data from parallel to serial,
- executes the modulation and demodulation needed, and
- performs external transceiver control.

The transceiver is an analog device which converts electrical and optical signals and uses the following parts:

- Digital-to-analog converter (DAC),
- Analog-to-digital converter (ADC),
- Preamplifier, and
- Gain-control circuitry.

In the protocol stack, each layer has certain responsibilities. The Physical Layer specifies the optical characteristics, encoding scheme and framing definition. The Link Access Protocol establishes the basic connection. An object- exchange protocol which enables the transfer of files and other objects. There is also serial and parallel port emulation and local area network access for notebook computers and other mobile devices.

The IrDA Control specification defines the transmission speeds, modulation schemes, and infrared (IR) wavelengths of the optical signals emitted by the transmitter as well as those signals received by the receiver.

An encoder and decoder reside in the bitstream path and handle data coding as well as the modulation process. Data are passed from the controlling device to the encoder/decoder and then on to the transceiver. A 1.5-MHz subcarrier process and the coding of the transmission signals are used to minimize interference with other transmission systems.

When an action is to be completed, the controller makes a decision and sends data out through the IR link to the other device. When polled by the host, the device responds, informing the host that it has information to send. The host then requests the information and the device sends it.

The controller passes the data on to the modem function, which handles the coding and modulation. The transceiver performs the electrical-to-optical translation between systems.

Two types of packets are used in the IrDA Control system short packets and long packets. The 1.5-MHz subcarrier pulses in the data bits are transmitted for a logical "1" and are not transmitted for a logical "0."

The host manages its communications with multiple peripherals on a time-division basis, using polled-response handshakes. The host polls all of the peripherals to determine which items need to be serviced.

The peripherals respond to the poll from the host, and do not initiate transmission. The peripheral devices do not transmit unless they are given response permission.

An exception is when the host is in sleep mode and a peripheral initiates a wake-up call for service. Then the host steps back into the polling sequence and looks for devices to service. If there is not any transmission between the host and any peripheral for a set time, the host again enters the sleep mode. Hosts do not communicate with each other. If multiple hosts are used, they detect each other and dither their transmissions to reduce the chance of interference.

The requirements for IrDA Control communication vary depending on the application. The sleep mode is used to minimize power consumption when a host and its peripherals do not need to communicate. The normal operational mode supports peripherals that may have different bandwidth requirements.

Pagers

Alphanumeric pagers are used for short messages. They usually display a few words or the caller's number. For lengthier messages, you can plug a PCMCIA device into your notebook or PDA to receive wireless e-mail. PCMCIA pagers can serve as pocket-size pagers and they can be plugged into a notebook when you want to read and compose longer messages.

Wide-area RF equipment may utilize one-way and two-way data flow. One-way data flow is the usual technology that is used for digital electronic pagers. In a one-way system, a carrier service, such as SkyTel, transmits a message and an address (the pager ID) over the geographic area it serves.

The pager recognizes its ID, then receives and displays the incoming digital message. Two-way service is similar except that the receiving unit also sends data back to the carrier, which forwards the data to another computer on the LAN. There are also PCMCIA cards that can receive and hold pager calls.

Wireless RF modems can be used to connect portable computers to a LAN. A cellular modem or modem/adapter that uses analog audio signals to convey digital information is called a voice-band modem or an analog cellular modem.

Circuit Switching

This is the common type of telephone data communications, where two users or their sending and receiving modems have the exclusive and full use of a telephone circuit or cellular channel until the connection is released. Circuit switching modems can be analog or digital. Unlike circuit switching, with packet switching no circuit is used on a dedicated basis. Data packets are sent along any channel or line available at the time they are received.

Digital cellular modems use digital rather than analog signals to convey information. The advantage is that digital eliminates static, line noise and signal loss during hand-offs. These handoffs occurs when cellular-based calls are transferred from one cell site to another as the user roams. Roaming occurs when you use a cellular phone in a city other than the one in which you originally set up your account.

TDMA (Time Division Multiple Access) is a multiplexing method for digital cellular phones in which each user is assigned a time slot on a channel. Often, ten users are assigned to a channel.

Code Division Multiple Access (CDMA) is a multiplexing method for digital cellular phones that uses a spread-spectrum technique. All speech bits (or phone calls) are assigned codes that are encrypted, transmitted, decrypted, then reassembled. By spreading the signals across the same broad frequency spectrum and assigning each a unique code, CDMA allows more calls to occupy the same space as each cellular channel is shared.

RF Modems

Radio frequency modems operate in the radio band, which is the portion of the electromagnetic spectrum from 100 kHz to 20 GHz. Wide-area radio-frequency services use an RF modem, which transmits and receives data in packets 128 bytes in size.

The advantage of packet data is that it optimizes throughput even under adverse line conditions. The sending modem sends a series of packets, without waiting to see if each is received successfully. If a packet is corrupted by noise or interference, the receiver asks the sending modem to resend the corrupt packet. The sending modem will continue to sending new packets until it receives this request. It then resends the errant packet and resumes normal transmission. Since each packet is numbered, they can be resent quickly, even out of order and correctly reassembled by the receiving modem without warming the throughput.

The Intel Wireless Modem uses firmware that allows the modem to use standard AT modem commands. Other wireless modems may require special software to operate since the data packets are assembled and disassembled by the computer. AT-type RF modems can be used by any popular communications package since the packet assembly/disassembly (PAD) hardware is right in the modem.

Circuit-switched Cellular Technology

Voice-band cellular or circuit-switched cellular systems use cellular telephone networks, their land lines and microwave technology to make the connection. Cellular phones are part of the mobile radio-telephone system that transmits data over a range of geographic sites, called cells which use low-power transmitters and receivers. The cellular phone can transmit data instead of voice using the audio signals produced by a modem. It is similar to a connection with two land-line-based modems. But, you can remotely access your LAN or call other mobile computers. You substitute a cellular phone for its land-line-based version.

The equipment needed for cellular data transmission is a special type of modem. The cellular modem connects to a cellular phone. There are also cellular adapters, which connect to both the modem and the cellular phone.

Most cellular modems can also use land lines to transmit data and to send and receive fax images. These eliminate the need for a separate land-line-based modem.

Not all cellular phones, modems, and adapters are compatible. A cellular modem or adapter must be able to control the bus of the cellular phone. Some companies such as NEC and Mitsubishi use open specifications which are widely supported by various cellular modem and adapter vendors. Others are not quite as open, and their modems and cellular phones must be used together.

Cellular Problems

Cellular channels can have problems with signal quality. Land-based telephone lines offer more consistent quality with minimal noise.

When noise is encountered it is usually due to a faulty connection.

This noise can result in a lower-than-usual transmission rate. Cellular channels are vulnerable to noise and interference which can change in intensity and frequency. Throughput can drop off quickly and the error correction and speed-setting techniques used for land-line telephones will not help.

The throughput may be cut in half when a large number of data errors are corrected. Compared to the faster modems the throughput seems slow and more costly, especially when sending and receiving large files.

Cell Protocol

Several companies have developed special cellular protocols. Microcom created an extension to its MNP-10 protocol. Unlike modems, which try the highest speeds first and then settle for lower ones until a reliable connection is achieved, MNP-10 starts at the low end and works its way up to the highest reliable rate. This allows modems to connect faster. MNP-10 boosts performance by shrinking and expanding the size of a data block as line conditions degenerate or improve. ZyXEL has modems with its ZyCellular protocol. AT&T Paradyne uses ETC (enhanced throughput cellular).

Cellular Security

Cellular telephone traffic, including data from cellular modems, is easy to intercept with inexpensive equipment. In the past even Radio Shack sold receivers capable of receiving cellular transmissions. Federal laws now prevent the sale of such devices, but many units still remain in use. While the chances of interception may be small, they are still very real. Unless some form of encryption is used, cellular modem users need to be careful about what they send over a cellular phone.

CDPD

The advent of digital cellular phone systems provided a whole new class of wireless technology called CDPD (cellular digital packet data). CDPD is compelling businesses and consumers to convert from analog to digital cellular. CDPD works by transmitting 128-byte blocks of data over the cellular phone system, looking for unused time in the digital cellular channels network, which is typically idle about 30% of the time. During the idle period, CDPD modems send short bursts of data packets to the cellular network. A cellular adapter connects a standard modem to a cellular phone (via the RJ-11 jack) and enables it to be used for cellular communication.

Unlike voice communications, CDPD automatically includes error-detection and retransmission so that no data are lost. It also has data encryption to secure the transmissions. Another advantage over RF modems and circuit-switched cellular, is that CDPD modems can exchange data with a designated host or service provider or be configured to work with any land-line network and modem.

CDPD modems use the same circuitry as digital cellular phones and some can double as digital cellular phones. Some modems are designed to work with portable computers like the IBM ThinkPad. Others are more flexible and work with any PC system.

Satellite Technology

Satellite technology can connect almost any two points in the world. Satellite equipment for a satellite-based phone includes the dish, stand, and transceiver. Some models only transmit Teletype and voice. Satellite units are called VSATs (very small aperture terminals).

Wireless LANs

For local area wireless access between two machines, the hardware can be infrared or RF transceivers. Wireless local area networks (WLANs) may show significant growth. There are extensions for the 802.11 standard for Ethernet-speed WLANs. Data rates are 20 Mb/s and higher in the 5-GHz frequency range and 10 Mb/s in the 2.4-GHz band. The Wireless LAN Alliance (WLANA) is a not-for-profit consortium of WLAN vendors which has pushed for the standard. It removes most obstacles to wireless LAN interoperability.

The modulation scheme used for the 2.4-GHz band was developed by Lucent Technologies and Harris Corporation. It is known as complementary code keying (CCK). This technology works in a wide range of different environments, including offices, retail spaces, and warehouses.

At slower speeds, the system shifts into a fall-back mode making it possible to increase the coverage areas as it falls back to lower data rates. This works on direct-sequence, spread-spectrum (DSSS) technology.

This fall-back is similar to cell phones, when there is a hand-off from one base station to another as a signal gets weaker and then stronger. In a wireless LAN, you essentially talk to an access point somewhere in the building, connected to the wired network. As you get farther away from it, rather than losing connectivity, it downshifts to a lower data rate, keeping the connection.

In multipath environments there are reflections which cause a delay-spread issue. When two wireless units communicate with each other, there will be reflected signals that will arrive at the antenna at later times. The delay spread is measured in nanoseconds and determines how long the radio will wait for the main signal to arrive. The longer the delay spread, the more data that can be received and the greater the accuracy of the received signal. As you increase delay spread, you increase the complexity of the unit and have to add equalizers to handle the reflected signals.

Wireless LAN products may use different modulation schemes, however there is a maturity in the 2.4-GHz technology. In the future we will see products at 5-GHz that match the specifications and requirements seen presently at 2.4-GHz. At 2.4-GHz there is the advantage of being close enough to the 1900-MHz personal communications services (PCS) phone band to be able to share low-cost components like filters and VCOs.

In older units, there are between 200 and 400 components—including filters, resistors, capacitors, voltage-controlled oscillators (VCOs) and the integrated circuits (ICs) on a PCMCIA card. In newer units the component count is down to 75.

As the trend to a high-data-rate pipeline to the home begins to evolve, through cable modems and ADSL, the problem is that once the data gets to the home, there is no practical way to move it around. It goes into one PC, which is ample if there is only one user. But, there are almost 20 million homes in North America with multiple PCs. In most cases, these are not on a wired network. These could be wirelessly connected giving every user in the household access to that high-data-rate line. WLANs can make it possible for multiple users to use the same Internet service provider (ISP) account. The wireless LAN industry is maturing, has standards in place, and is moving toward the next evolution in products.

Pocket Communications

Wireless systems devices are evolving as hand-held, multifunctional, portable communicators combine the capabilities of digital cellular phones with personal digital assistants (PDAs). The latest designs integrate features from multiple communications products including digital cell phones, pagers, data/fax modems, voice messaging and high-end PDAs into a single hand-held unit. Each of these functions is typically handled by separate signal-processing integrated circuits (ICs) called digital-signal-processors (DSPs).

Integrating these functions into a single DSP requires signal-processing performance beyond the capabilities of traditional DSPs which use DSP assembly code and required highly skilled, specialized programmers. This time-consuming and expensive method results in code that is difficult to maintain.

New more-efficient DSP architectures have appeared which use techniques and architectures from high-performance microprocessors. Some examples are superscalar/reduced-instruction-set-computer (RISC) and very-long-instruction-word (VLIW) architectures. Both of these can provide 10 times the performance of traditional DSPs. These DSPs have the following characteristics:

- wired functions data/fax modem,
- wireless functions cellular phone, paging, and voice messaging,
- telephony functions telephone-answering with digital recording and playback.

The development of wireless solutions combines integrated-circuit technology advancements with system-level analysis of the desired application. Innovative IC designs following a continually improving process technology allow wireless manufacturers to provide cost-effective products. Advancements in ICs for wireless applications include:

- faster digital-signal-processors (DSPs) with more million instructions per second (MIPS),
- programmability,

- increased levels of integration within the DSP,
- larger amounts of on-chip memory,
- intelligent input/output (I/O) peripherals and,
- coprocessor for accelerating specialized functions.

A cellular base transceiver station (BTS) serves as the interface between mobile terminals and the rest of the cellular network. The frequency spectrum is allocated in 200-kHz channels, with each channel's bandwidth supporting up to eight users concurrently.

The BTS contains one transceiver for each channel that is allocated to the cell. The BTS must send, receive, and process all of the cellular transmissions in the cell. Besides the speech or data included in the cellular transmissions, timing, synchronization, and control data are used to make the cellular network function nearly seamlessly and transparently to the cell phone users as they roam among the cells.

The size of a cell depends on the terrain and the amount of calling traffic. Cell sizes range from a maximum radius of approximately 35 km (macrocells) to a few kilometers (microcells) or several hundred meters (picocells). Picocells are needed in densely populated areas.

In a channel, a time division multiple access (TDMA) format is used so that all users share the same bandwidth. The TDMA has time slots for eight users. The cellular data are transmitted in bursts and each burst fits into an allocated time slot.

The base-station controller (BSC) controls BTS functions for several cells. It hands off communications from one BTS to another as a communicating mobile terminal travels from cell to cell. It also performs speech coding and rate adoption to seamlessly connect the calls between the cellular network and the public-switched (land-based) network.

The transmit function involves channel encoding, interleaving, and encryption. The receive path must perform equalization, deinterleaving, channel decoding, and decryption.

Speech Coding

In the transmit path, the digitized speech is first encoded. Speech encoding compresses the speech through processing that uses speech modeling and prediction. Speech encoding also detects pauses in speech and ensures that the data generated during the pauses will not be transmitted.

The channel encoder packs the data of each speech frame with error-correction information so that the data are immune to the interferences that can occur when signals are transmitted over the air.

The data are encrypted to ensure security during cellular transmissions. Encryption uses the XORing of the data bits with a key sequence generated by an encryption algorithm.

Equalization is used because of the multipath effects of the received signal. A transmitted signal may take multiple paths to the receiver antenna which creates multiple versions of the same signal. Each of these arrives at slightly different times and with different signal strengths. An error-correction coprocessor operates in parallel with the DSP and accelerates the equalization and the channel coding.

To assist the receiver in detecting the true signal from the multiple versions, the transmitter encodes a training sequence into the burst. The receiver's equalizer knows the training sequence and uses it to identify the transmitted signal.

The trend is to integrate the DSP with analog-to-digital converters (ADCs), digital-to-analog converters (DACs), a clock synthesizer, and system control units that include a memory controller, an interrupt controller, and serial I/O controllers. The functionality that is usually integrated across an entire chip set is created on a single chip.

The BTS must be software upgradeable in the field so that it can be easily reconfigured to support evolving standards over its life cycle. On-chip flash memory is used so that application software can be loaded and debugged on the spot.

The protocols of narrowband personal communication services (N-PCS) include Motorola's FLEX. The N-PCS frequencies lie in narrow bandwidths just above 900 MHz. Some of this spectrum is allocated to 12.5-kHz paging response channels, allowing pagers to transmit replies to received messages. Four-level frequency shift keying (FSK) modulation is used by FLEX. A synchronous protocol allows the pager to switch on periodically (every few minutes) and check if a message is being sent.

Decoding starts with frame synchronization where the decoder aligns its receive buffer to the frames in the received bit string. Data decoding algorithms allow the extraction of message data, error-control data, and ID data from the frames. The error control code is used to correct errors in the frame data. In some protocols, such as the European Radio Messaging System (ERMES) and the Motorola FLEX paging protocol, deinterleaving is also performed.

The address field encoded into the frames determines if the message is for the particular pager. If not, the decoder informs the controller and the controller conserves power by shutting off parts of the receiver. The message data may be passed to the controller, which then delivers the information to the user either by a liquid-crystal display (LCD) or speaker.

Wireless and wired functions are similar in some ways and in other ways they are not. Sending data over a wireless channel poses some problems due to the channel's corruptive nature, compared to data transmission over phone lines, which offers a relatively clean channel.

Compared to voice, data transmission via a wireless channel has two added complexities every bit is equally significant and must be recovered at the receiver, and the data rates are much higher.

The multifunctional communicator's DSP system must be capable of on-the-fly reconfiguration. The multiband and multimode capabilities of cellular-phones require the DSP to reconfigure the phone depending on the user's location. There are different carriers and standards that may be available locally. The software must reconfigure the unit to handle a function like narrowband cellular phone, digital audio receiver, or wideband high-speed data transfer.

Multiband capability means that the cell phone must be able to perform a search algorithm that determines the available air interface. Multimode capability determines the carrier, cellular standard, and quality level available.

The DSP also performs filtering that was previously handled by the RF transmitter/receiver. This allows the DSP to control the operating frequency band of the unit. This filtering takes large amounts of data at high speed, which requires a high DSP throughput.

One of the algorithms used in digital cellular phones is a speech codec. This algorithm takes digitized speech and compresses it to a lower bit rate to conserve transmission bandwidth. Another is error correction. The error protection and correction technique is called convolutional encoding and decoding, which adds redundancy and structure to the compressed voice data to correct for bit errors. Convolutional decoding uses the Viterbi algorithm to recover data bits from the received data. In asymmetrical-digital-subscriber-line (ADSL) systems data rates are so high that a Reed-Solomon algorithm is combined with bit interleaving and a cyclic redundancy check (CRC) to protect and recover data.

Equalization is used in wireless receivers to clean up the received data stream of digital data. Transmitting data in a mobile environment will corrupt the transmitted signal due to multipath and Doppler frequency shifts. An equalizer uses the Viterbi algorithm to reverse these effects and resynchronize the bit stream.

Next-generation DSPs are enabling the next generation of PDAs by providing the horsepower required to implement multiple tasks on a single device. Quality voice storage and playback are part of the PDA of the future. The DSP can compress, store, and reconstruct speech. If the DSP is performing speech coding, speech samples are transferred between the DSP and the voice-band codec.

Since the system must be reconfigurable, it must be programmable and capable of handling multiple tasks simultaneously and shifting to new tasks as needed. This means the DSP must offer flexible multitasking support for the simultaneous operation of functions.

The PDA of the future uses a group of DSP algorithms that are dynamically executed. The development of these algorithms usually begins with a high-level-language (HLL) model. A compiler allows the porting of the model to assembly language with some hand-coding support to optimize the final program.

Next-generation DSP architectures are based on superscalar machines to provide the required throughput needed. Superscalar DSPs use multiple execution units coupled with high data bandwidth to execute many instructions per cycle.

Superscalar DSPs are scaleable since the architecture can be tailored to an application by changing the number or type of execution units to fit particular needs. Superscalar devices will continue to evolve to address the continuing demands of wireless applications.

Data Exchange

Personal digital assistants (PDAs) along with pagers and smart cellular telephones can share Windows data. There have been a number of advances in this exploding world of miniature, mobile electronics.

Apple's Newton MessagePad and the Sharp Expert Pad are not wed to any one operating system. The Windows version of the Newton Connection Kit lets you transfer data between this infrared-equipped PDA and Windows systems. Some PCMCIA-based Newton applications are also available in Windows versions so you can run the same software on both machines.

The first PDAs from Apple, Sharp, Tandy and Casio were designed as computing devices, with communications available as an option. Later devices like the Simon cellular telephone were designed as cell phones with computational functions added in. Developed by IBM, the Simon used a tap-sensitive LCD screen that you use to access built-in features or to dial a number. Most PDAs prefer PCMCIA interfaces.

BellSouth's Simon cellular phone uses a predictive keypad. Instead of cramming a full set of tiny characters on the screen, it displays on a pen-input screen only the characters most likely to follow the previous character you entered. If it guesses incorrectly, you hit an icon to call up another set of possibilities.

Computers are getting more portable with units like Xybernault's wearable PC. It has a 233-MHz processor, 128 MB of RAM and a 4.3 GB hard drive. The 28-ounce processor units attaches to your belt and the keyboard can be strapped to your arm. Qual Comm's PdQ Smartphone combines a 3 Com Palm III and a Qual Comm digital cellular phone. You can dial directly from the Palm III address book.

Internet phones are another developing area. PC-based Internet phones use programs that let you talk to others on the Internet. You need a sound card, speakers, microphone and modem. IDT's Net 2 Phone is a program that connects your computer to a real phone.

New generations of digital cellular modems will use the entire channel for data and connect with any other modem. These modems using digital technology promise to be more immune to noise and to offer higher data throughput rates.

Third-generation Wireless

Third-generation wireless phones will drive the convergency of computing while creating a range of new wideband wireless multimedia products that are capable of delivering not only voice, but also video and data-intensive information, such as real-time Internet access.

QUALCOMM has been involved in next-generation mobile phone standards which would integrate the well-entrenched Global System for Mobile Communications (GSM) network with a standard known as cdmaOne, which depends on code-division multiple access (CDMA). Ericsson has been working with Japan Telecom to develop wideband CDMA (WCDMA) as an enabling technology for third-generation services. These new, emerging services may be developed for time-division-multiple-access (TDMA), as well as GSM and Japan's Personal Digital Cellular (PDC) service. Vodaphone, which is a major United Kingdom mobile phone service provider has conducted field trials of a system that integrates the GSM network with the cdmaOne interface. Software-definable radios are another possible approach.

While there is not a world phone, there is growth in easy-to-upgrade software-based handsets. As you move around the world, you would load different software to adapt to the phone systems you want to use. This still requires huge doses of integration and computer power.

Services based on the Universal Mobile Telecommunications Standard Institute (UMTS) should be running in the 2-GHz band. UMTS should deliver wide-area, high-mobility data rates of 384 kb/s and local-area, low-mobility rates up to 2 Mb/s.

Almost 240 operators in over 100 countries operate GSM so Ericsson, Nokia, and Siemens are all committed to the standardization of UMTS based on GSM.

The focus of UMTS has shifted from broadband integrated-services digital network (ISDN) to the Internet. The first phase of UMTS should have voice services augmented by data services at 144 to 512 kb/s.

IMT-2000 is Europe's proposed system that would integrate wireless systems, including satellites. It would optimize the use of the Internet. Japan has its own version of IMT-2000.

IMT-2000 services include high-quality voice/audio, high-speed data transmission with still images, video conferencing and multimedia. Subscribers to IMT-2000 will be able to carry on voice conversations while accessing a corporate Intranet, or take part in video conferences while sending and/or receiving e-mail or data.

Mobile Internet

Mobile is posed to become a major access medium for Internet and on-line services. Web browsers for cellular phones and personal digital assistants (PDAs) are in use and thriving. Internet paging is also growing rapidly.

Analog and digital cellular systems already provide circuit-switched Internet access. Much faster direct digital connectivity will be available as multislot transmission over GSM networks. Another high-speed data transmission technique for GSM is General Packet Radio Services (GPRS). It is currently working its way through the standardization process. In time-division-multi-access (TDMA) networks, cellular digital packet data (CDPD) can also be used to provide Internet access.

High-speed circuit-switched data (HSCSD) will allow the downloading of graphics-heavy pages from the World Wide Web (www) to be as simple over mobile as it is over terrestrial services. GPRS is perfect for e-mail and data access services, where users do not want to pay high call charges for short transmissions. GPRS also allows the user to receive voice calls simultaneously when sending or receiving data calls.

HSCSD has built-in bandwidth-on-demand capability. So, it will not matter what speed the user needs for any particular application, the service will provide whatever speed is required, up to 64 kb/s.

Another emerging service is voice-over-IP, or Internet telephony. Voice-over-IP is generally delivered from one personal computer to another through modems and special software. Instead of running on general-purpose processors, improved voice quality is possible using digital-signal-processors (DSPs).

Bellcore's Mobile Network Computing Protocol (MNCP) for the Internet community may become an open standard. It is a generic protocol specification that allows any cellular, personal-communications-services (PCS), or paging network to provide wireless Internet information access from mobile computers and hand-held devices.

This standard would allow portable and accessible mobile computing devices to offer easy desktop connectivity and allow users to remotely perform computer operations such as control system management.

In the future networks and mobile phones will evolve in parallel to provide services that are increasingly personalized to the users' needs. This will happen as a result of increasing intelligence within the networks along with increasing competition and segmentation for services.

Smart Antennas

Smart antennas are becoming integral elements of evolving analog and digital cellular networks. A smart antenna system is a highly configurable antenna system that senses the radio environment, including traffic and interference levels, and then modifies its operation to optimize performance for local conditions.

The intelligence of the antenna system resides in the algorithms that monitor the RF environment and adjust the antenna's radiation pattern in order to increase carrier-to-interference ratios, reduce network-interference levels, control hand-off boundaries or redistribute traffic loads.

Smart antennas will be used with code division multiple access (CDMA) technology in mobile cellular and personal-communication-services (PCS) networks as well as fixed-terminal-wireless-local-loop systems. Smart antennas are emerging as a major component of wideband CDMA standards for third-generation mobile systems in North America, Europe, and Asia.

The technology of smart antennas evolved from phased-array techniques used in military and radar applications. These advanced electronics concepts are now practical for wireless communications networks. A smart antenna platform uses a phase-array antenna to synthesize custom antenna patterns.

In Advanced Mobile Phone System (AMPS) networks, smart antennas create narrowbeam antenna patterns that track subscribers as they move around the cell site. The narrow antenna beams improve call quality. Smart antennas can provide network-wide interference control, traffic-load balancing, resource allocation and dynamic network management.

In the CDMA operating mode, the smart antenna uses a phased array to create custom sector-antenna patterns through a process called sector synthesis. This is done with software management. Sector synthesis controls the azimuth angles, beamwidths and gains.

Service providers can match antenna patterns for the different sectors of a CDMA cell. The smart antenna system monitors traffic loading and interference levels on the CDMA links. The smart system then responds to the traffic load and interference levels by creating antenna patterns that are used to equalize traffic loads and reduce interference.

In cellular networks, traffic loads are unevenly distributed across cells and sectors. A cell may have one sector near the blocking point, while the cell's other two sectors are lightly loaded.

Traffic load balancing is used to shift the traffic load from heavily loaded sectors to underused sectors. This results in a reduction in the peak loading levels and allows an increase in the traffic or network capacity. Without traffic load balancing, only about 1/2 of the subscribers in the highly loaded spot receive acceptable service.

Smart antennas also allow rotation of the azimuth angle of the sectors in order to shift hand-off boundaries. A feature called soft hand-off allows subscriber units to be simultaneously connected to multiple sectors. This results in an increase in the total average transmit power per subscriber, which wastes valuable resources.

Excessive hand-off activity can also result in dropped calls due to hand-off failures in areas where there is no dominant server. Smart antennas make it possible to synthesize radiation patterns that reduce hand-off problems and maintain coverage.

Varying interference levels across the network mean that coverage, quality and capacity change based on geography and time of day. During network installation and network maintenance, a significant amount of time and effort is used to fine-tune interference levels. Operators adjust transmit powers, change antenna patterns and tweak network parameters to eliminate interference in problem areas. Smart antennas provide exceptional flexibility in tuning the RF footprint of each sector. Transmit power can be turned up in certain directions to enhance in-building coverage. In other directions, transmit power can be reduced to minimize interference, control handoffs and reduce coverage overshoot.

The flexibility of smart antennas allows service providers to customize RF coverage footprints to precisely manage interference, hand-off levels and traffic load balancing. Sensing the radio environment and providing a flexible way of responding to changes allows smart antennas to play a leading role in building communications management.

ISDN

The Integrated Services Digital Network (ISDN) is based on switched digital technology. It can be used to upgrade analog telephone systems to an all-digital end-to-end network. ISDN can be used for voice, data, video, fax, and image transmission. ISDN fills a gap between slower switched analog (modem) connections and dedicated digital (leased-line) connections. ISDN is both switched and digital. A switched connection lets anyone on a network make an on-demand connection from one site to another.

ISDN is all digital, so with computer-generated information already in digital form, you do not have to convert computer data from digital to analog form and convert it back to digital on the other end. ISDN has a low bit-error rate compared with modem transmissions and you also have the advantage of virtual connections. With a virtual connection, you can dial into an ISDN bridge or router from a remote site and log in to a file server. While connected, you may get a phone call, or some other distraction, and stop work on your computer. Since the connection is idle for a few minutes, the equipment could drop the line while maintaining your log-in to the file server. It does this by spoofing the protocols.

When you later require access to the file server, the ISDN equipment re-establishes the connection, typically in less than one second. This is very fast compared to the time required to establish a connection between two modems.

ISDN equipment is termed customer premises equipment. There is a basic rate interface and a primary rate interface. The basic rate interface (BRI) uses the telephone company's existing two-wire copper lines and multiplexes three channels over them. The primary rate interface (PRI) has 23 B channels and one D channel in the United States while 30 B channels and one D channel are used in Europe. In the United States, PRI runs on a T-1 local-loop circuit. In Europe, an E-1 circuit is used.

The B channel can be used to transmit voice, data, video, image, or fax. It can operate as circuit switched voice (CSV), circuit switched data (CSD), or packet switched data (PSD) which is also called packet mode data, or PMD. It runs at 56 Kbps or 64 Kbps, depending on the switching mode and on whether the telephone system can handle clear-channel data calls.

The D channel provides call-setup times of about one second. The D channel can also be used for packet-switched data. The ISDN equipment can use the traffic between sites to add additional B channels to increase the throughput. The number of B channels used can fluctuate up and down throughout the day, providing increased bandwidth at peak times and dropping back during low traffic periods to reduce phone charges.

ISDN equipment can also compress data with compression ratios comparable to those of modems. Typically, ISDN achieves compression ratios of about 2:1, but 4:1 or even 8:1 are possible under ideal conditions.

ISDN gives twice as many phone lines (B channels) per number of copper-wire pairs as analog, with packet-mode data on the D channel also being available. With standard four-wire copper cable, analog telephones and modems, you get a maximum of two telephone lines. ISDN users can have four numbers on these same four wires along with two additional 9.6-Kbps packet-mode connections.

Internet Connections

One of the major uses of modems is Internet access. The Internet consists of local connections, long-haul connections, routers, servers, and a variety of protocols. These components work together in the following ways.

Users connect by dialing in to an Internet Service Provider (ISP) using a modem over telephone lines. These are often the same kind of twisted pair, copper wires that have been used since the telephone was invented. At the ISP end is another conventional modem. These modems convert the computer's digital information to analog signals and then back to digital on the remote end. Connection speed is at best, not counting factors like line noise, the speed of the slower of the modems.

If you are connected to the Internet through a network, your PC is part of a local area network (LAN). Your PC has a network interface card (NIC) that ties it to the network which is typically Ethernet.

The cable coming out of the NIC is either coaxial cable or a larger version of your telephone wire. Your network will be probably connected to an ISP not by a modem, but through a higher-speed connection leased from the local phone company. Newer technologies offer high-speed connections to users.

Internet Growth

The Internet was born over decades ago in the form of a U.S. Defense Department network called the ARPAnet. This was an experimental network designed to support military research in communication networks that could withstand partial outages from military operations and still function.

In the late 1960s most computers were not capable of sending graphics across networks or even viewing them. The traffic that traversed the network was generally text-based in the form of electronic mail messages and the direct interconnection of one computer with another.

In the 1970s the Internet began to be interconnected with large universities and research organizations. The nature of the traffic began to move towards on-line research, but it was still government-oriented.

To send a message on the network, a user only had to put the data in an envelope, called an Internet Protocol (IP) packet, and address the packets correctly. The communicating computers were also given the responsibility to ensure that the communication was completed.

By the late 1970s, there was a need to standardize the way computers on the Internet communicated with each other. This resulted in TCP/IP. TCP/IP stands for Transmission Control Protocol and Internet Protocol. These networking protocols allow different types of computers to communicate.

In the mid-1980s, the government became concerned about the security of defense information and split the network into ARPAnet and MilNet. The responsibility of ARPAnet was moved over to the National Science Foundation (NSF), which merged ARPAnet into the high-speed network called NSFnet.

NFSnet became the prototype for national backbones. The high-speed links were T1 phone lines which served as the main traffic arteries across the country. Network traffic increased until the computers controlling the network and the telephone lines connecting them were overloaded. In 1987 the old network was replaced with faster telephone lines and faster computers to control it.

The system began to be used for a wider variety of data and the national Internet emerged. In 1992 the NSFnet backbone was upgraded to higher-speed T3 links, which operated at a speed of 45 Mbps.

In 1995 NSF released control of the Internet backbone to the commercial carriers. Since then, NSF has focused on providing regional Internet connections and building a faster backbone. The Internet grew and became a practical method of communication for different computers from various sites and companies.

The Internet is the largest network we have, with all types of different computers connected to it. These computers talk to each other using protocols, which set the rules of communication. Without a common set of rules the different computers would not be able to send and receive information among themselves.

The protocols that make up the Internet are the Transmission Control Protocol (TCP) and Internet Protocol (IP). TCP breaks down information into packets or envelopes of data for transfer and reassembles them at the destination point. IP ensures the proper delivery of data to the right address.

Any computer that wants to communicate with any other computer on the Internet must use TCP/IP. Associated with the TCP/IP protocol are several TCP/IP applications or tools for sending and receiving electronic mail, transferring files, and accessing files remotely.

TCP/IP is a packet-switching networking scheme. Information is sent across the network in a packet that contains source and destination information. Since each packet has its addressing information built in, it can travel independently. This allows large numbers of packets to flow through the network taking different routes. Packets may arrive out of order, but because each packet also contains sequence information, the receiving computer can reconstruct the original data.

The high-speed central networks in the Internet are known as backbones. They are like a superhighway system and they deliver data to mid-level networks. The mid-level networks take the traffic from the backbones and distribute it to regional networks which are more like local highways in the automobile world.

Client/Server Systems

Any two computers connected on the Internet act like they are in a client/server mode. A client is a computer connected to a host computer, which is called a server. The client software is used to run programs across the network. The client program provides the user interface for working with the server. It takes commands you execute from your computer and transfers them to the server program running on the host computer. For example, a client program called browsers allows you to connect and work with a World Wide Web server.

Although the Internet uses the telephone system, it is not a network primarily based on dedicated telephone lines. The telephone network is a circuit-switched network. When you make a call, you get a switched or dedicated connection. Even if you are not talking or listening, this connection is unavailable to others who may want to use it. This means there is some under utilization of this resource. The Internet works more like a mail service where mail is mixed together and sent in a bag or packet to another location and sorted out for its destination.

Routers

The Internet uses computers called routers to connect the various networks together. These can be LANs (Local Area Networks), regional networks and even telephone lines. Information on the Internet moves through the telephone lines and the other nets. The routers make decisions on how to route the data packets so they reach their destination.

Each router needs to know what connections are available for the next hop to get the packet closer to its destination.

The routers are like the hubs of the network. Each router connects several different networks. The router calculates the best routes for the packets to take and moves the packets through the different networks.

The Internet Protocol (IP) takes care of addressing, so the routers know what to do with the data packet when it arrives. The Internet relies on its routers to get data to the right network, and then that network gets the data to the right computer, or host, on the network.

Addressing

Each computer connected to the Internet must have its own unique address. The addressing information goes at the beginning of the message. This gives the network enough information to deliver the packet of data.

The two main types of addresses on the Internet are IP addresses and domain names. Each computer that uses TCP/IP protocols is distinguished from other computers on the Internet by a unique IP address. The IP address is made up of four numbers separated by periods. These four areas of the IP address are hierarchical.

Each number is less than 256. The beginning of the address tells the Internet routers what network you are part of. The last part of the address tells that network which computer or host should receive the packet.

Originally, the Internet used only IP numeric addresses, but these proved cumbersome to remember, so a text system was developed called the domain name system (DNS). These domain names indicate the name and type of service or organization that conducts the service as shown below:

Domain	Organization
com	Commercial and industrial organizations
edu	Education (universities and other educational organizations)
gov	government (non-military)
org	Other organizations (nonprofit and research organizations)
net	Network operation and service organizations

As long as the data are placed in an IP envelope, the network has the information it needs to get your packet from your computer to its destination. The information in a packet is usually between 1 and about 1500 characters long. This prevents any one user of the network from monopolizing the network. It also means that the network becomes slower as more people try to use it.

Many information transfers are longer than 1500 characters and networks sometimes lose packets, or damage them in transit. Packets can also arrive out of sequence. The next layer of the network provides a way to transfer longer messages and also takes care of some of the degradation that can occur in the network.

Transmission Control Protocol (TCP)

TCP is the protocol that handles most of these problems. It makes sure the packets arrive and pastes them together in the right order. TCP takes the information you want to transmit and breaks it into sections. Each section is numbered so receipt can be verified and the data can be put back in the proper order. The sequence number is sent in a packet of its own. Some or all of your data are placed in a TCP packet.

During transmission, packets can be lost, or affected by noise on telephone lines. TCP handles this by calculating a checksum. This is a number based on the parity of the transmitted bytes. Parity involves counting the number of 1s and determining if this number is even (even parity) or odd (odd parity).

When the packet arrives at its destination, the receiving TCP software calculates what the checksum should be and compares it to the one sent by transmitter. If they do not match, an error has occurred during the transmission. The TCP software throws out that packet and requests a retransmission.

The TCP packet is placed inside an IP packet and given to the network for transmission. On the receiving end, the TCP software collects the packets, extracts the data, and puts it in the proper order. If some of the information is missing, it asks the sender to retransmit them. When it has all the information in the proper order, it passes the data to the application program that is using its services.

TCP appears to the user as a dedicated wire between two applications. There is not a dedicated link between the sender and receiver since others will use the same routers and network wires in their transmission.

Dial-up Internet Access

The easy and most popular way to get network access is to get an account on a computer that already has dedicated access. Then you use your computer to log in to this system. Dial-up access is not as good as having your own connection, but it is easier and less costly to set up. Your computer does not actually become part of the Internet. It accesses a service computer from an Internet service provider that is permanently connected to the network. There are many organizations that provide this type of service now. They are generally known as Internet Service Providers (ISPs).

Future Internet Service

A significant number of Internet users will access the net over cable or DSL connections that are up to 50 times faster than 56-kbps modems. Many people will use the Internet for video conferencing, telephony, telecommuting, and on-line control monitoring applications. These applications demand raw bandwidth and a level of reliability the present Internet cannot provide. In the near future the Net will have to carry more data than it does today and with more reliability.

At 56 kbps, today's modems are going about as fast as today's phone lines will let them. Access speeds vary from 28.8 kbps, or slower, for older modems, to 56 kbps for newer models and for some users, 256 kbps to 1.5 mbps for DSL, and up to 3 mbps for cable modems. There are a variety of broadband options, including T1 lines and other dedicated, high-speed connections (Table 9-4).

Cable and DSL are two technologies trying to replace traditional modem connections. Current cable connections can manage speeds up to 3 mbps. DSL implementations run between 256 kbps and 1.5 mbps.

Another option is Sprint's Integrated On-demand Network (ION) which can deliver voice, video, and data at speeds up to 620 mbps. An ION Integrated Service Hub is wired into your phone jack and attached to an Ethernet card in the PC.

Table 9-4. Internet Services

Service Type	*Availability*	*Typical Performance (downstream/ upstream)*	*Features (pros and cons)*
Dial-up	Universal	56 kbps/33.6 kbps	Cheap, easy to install, and readily available. Slow service ties up a phone line.
ISDN	Widespread	128 kbps/128 kbps	Twice speed 56-kbps modems plus phone service features like cell handling. Costly.
Satellite	Widespread	400 kbps/33.6-kbps	Good downstream speed, need clear view of southern sky. Uses dial-up for upstream connections, complex installation.
Cable	Limited	1-5 mbps/33.6 kbps 2.5 mbps	Where available, inexpensive continuous high-speed connection. Cannot select ISP. Shared nodes can cause security and speed problems.
DSL	Very Limited	144 kbps-8 mbps/ 64 kbps-8 mbps	Turns your phone line into a fast connection that is always on.
Frame Relay & T1/T3	Widespread	56 kbps-45 mbps/ 56 kbps-45 mbps	Business-oriented/speed guarantees/quick repair. Costly, complex hardware.

Some companies like Teligent and Winstar offer transmission using wireless radio frequencies instead of congested copper lines. Voice or data would be transmitted from a small dish antenna on the roof to a central office, which would then transmit the information to your ISP. Designed primarily for urban areas, this arrangement could be a lot cheaper than conventional cables.

The Future of ISP

As end users and businesses move up to faster connections, the bottleneck may shift to the ISPs. The problem is one of capacity. If large numbers of users with access to 1 mbps. DSL signs up for the service, the infrastructure would probably be able to handle it. But, this is not true for 7-mbps DSL and 20- to 25-mpbs DSL will be available in the near future. The network will not be able to handle this traffic.

In addition to selling access to end users and businesses, large national ISPs such as Sprint, MCI WorldCom, and AT&T sell bandwidth to smaller ISPs. Those smaller ISPs connect at Network Access Points (or NAPs), which in turn connect them to the backbones (See Figure 9-2).

Few businesses have cable since the cable companies did not wire business and industrial areas. But, they do have access to high-speed dedicated lines ranging from 56-kbps frame relay to 1.5-mbps T1 lines and beyond.

The Internet backbone is really several backbones today. The big national ISPs, along with several smaller firms, own high-speed fiber-optic networks that connect major metropolitan areas at speeds up to 38 gbps. In a few years, those connections could be as fast at 200 gbps.

Besides T1 or frame relay lines, other technologies include the Integrated Services Digital Network or satellite-dish access that can provide even faster access.

For many, the future lies in cable modems and Digital Subscriber Lines. Both cable and DSL are much faster than analog modems. Each provides a reliable connection and fast connect times.

At the end of 1998, the number of cable modems in use in North America was about 513,000 and the number of DSL lines in the

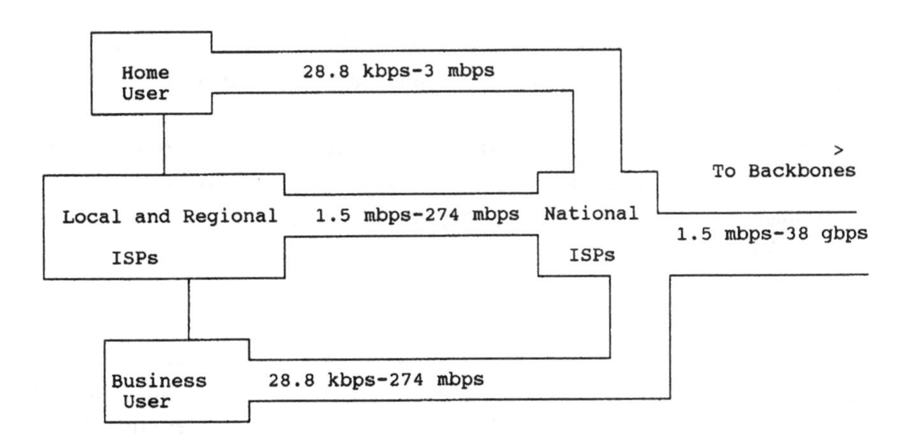

Figure 9-2. Internet connections.

United States was just 39,000. In contrast, there are tens of millions of consumers around the world connected to the Internet via dial-up modems.

A lack of standards has impeded wider cable and DSL availability, but that should change soon. Growth is projected at 1000% or more per year for both technologies in the near future.

Dial-up connections are cheap but slow. Frame relay and T1/T3 lines are fast but expensive. Alternatives include speedy, affordable DSL services or cable with fast performance and even lower prices.

While cable TV lines are available to more than 100 million North American homes, only 20% of these can get cable modem service. That number will grow as cable companies offer two-way fiberoptic networks that deliver more TV channels as well as high-speed Internet connections.

Cable services can deliver connection speeds at least 10 times faster than a 56-kbps modem for about twice the price of a typical dial-up account.

Getting a cable modem requires a PC with an Ethernet card, which the cable company can supply and install. The cable company will often lease the modem as part of the monthly fee. Most cable systems offer always-on service that does not tie up a phone line. About 15% use telco return, which requires a standard dial-up line to send data from your PC to the Internet.

A Web page that takes several minutes to download with a 56-

kbps dial-up modem could be done in under 5 seconds with a cable modem. Most cable modems can speed up file transfers by 250 to 1000%.

Cable modem service cannot abate Internet traffic that occurs outside its own network. But, many services cache recently visited Web sites on their own servers, so popular pages and files can download more quickly.

Cable performance can match and sometimes beat that of more expensive T1 lines. Cable modem service can deliver cheaper speed in part because it does not give you a dedicated fast connection to the Internet. It divides the total cable which may be capable of transfer rates up to 36 mbps between users on your node. Depending on your service provider and your location, you might share your connection with hundreds or even thousands of other users.

If too many users on your node hit the Internet simultaneously, your connection could suffer. Some cable modem users reported incidents where performance dropped off to 2 or 3 kbps. Cable modems can add more capacity. They have the architecture to allow higher capacity.

Internet2

Internet2 will connect only academic and government users. The rest of the Internet will benefit indirectly as high-bandwidth Internet2 applications trickle down. There will be greatly improved multimedia and video conferencing capabilities. Internet2 groups are developing techniques for tagging voice data packets so they get a higher priority as they pass through the network.

Another promising development is occurring in a branch of physics called photonics. Today, information travels down fiberoptic cables in the form of laser-generated light. Photonics involves ways of handling these light waves to increase the fiber's transmission capabilities.

One technique is Dense Wave Division Multiplexing (DWDM). It sends multiple frequencies of light down the same fiber at the same time. Sprint uses DWDM to increase the company's fiber capacity by a factor of 32.

In the future it may be able to increase capacity by a factor of

100 or more. Some cable companies are using AT&T's DWDM-based backbone to carry data traffic.

SilkRoad has another photonics technology called Refractive Synchronization Communication (RSC). It has demonstrated data transfer rates of 200 gbps over 200 miles, without any signal amplification, on a single wavelength of light. The fastest fiber in use today has a data rate of 40 gbps. SilkRoad has been negotiating with local telephone companies and larger carriers that could use RSC technology soon. Expect to see local area network RSC-enabled hubs and switches used by businesses and homes that have fiber connections. The entire Internet backbone could be using photonics in the near future.

Internet Bottlenecks

One segment of the Internet could accelerate its data rate, but the others must struggle to keep up. Now, the bottleneck is at the desktop. As cable and DSL use grows, the ISPs and backbone providers will have to speed up their segments to accommodate the new traffic.

The telecom industry is in the middle of a transition from circuit switching to an advanced public network. Service providers are progressively building a packet-switched and cell-based infrastructure that can handle data, video and voice. The transactional network will use a jumble of gateways to connect legacy circuit switching with cells and packet switching.

The new public network will have a packet and cell-based infrastructure that will be equally adept at handling data, video, and voice. When this convergence is complete, it will change everything.

Instead of buying pipes for bandwidth, you will be able to sign up for a new class of VPNs (virtual private networks) that will run over managed backbones. These will be tuned to deliver different types of traffic. Cheaper bandwidth will be offered. The new technology should also allow 1-Mbit/s ADSL modems. Bandwidth will increase in the same way that memory has increased in PCs.

Data traffic is growing at a remarkable rate. Today, there are about 15 million fiber circuits in North America and 120 million copper lines. Creating the new public network means most of the equipment needs to be upgraded from the central offices to the infrastructure.

Conventional circuit switching will not do in the 21st century and boosting bandwidth may require gear like dense wavelength-division multiplexers (DWDMs). The new technologies also include terabit routers.

Another factor is the Internet, which dumps a huge amount of data onto public networks. Data on some backbones is growing at 800% annually. In 1996, international Internet traffic was so low compared with voice that it was scarcely a ripple, but in only 18 months it rocketed up to exceed voice.

This data explosion is forcing carriers to curb their investment in circuit-switched equipment. The Internet is also forcing carriers to standardize on IP (Internet Protocol).

The old public network was for voice. Everything is standardized on 64 kbit/s voice channels. Voice does not need 64-kbit/s any more, it can be compressed without sacrificing quality. Trying to accommodate bursty packet data on fixed circuits is fundamentally inefficient. Most capacity remains idle much of the time.

Each data service runs over lots of overlaid networks. This results in inflated service charges. In the new public network, traffic will be streamed into a single fiber that can handle up to 80 Gbit/s. There are already predictions of even faster optical switches.

New types of equipment include monster RASs (remote access servers) that terminate thousands of analog calls from analog modems. The new public network is supposed to be packet-or cell-switched. But, the old public net will be part of the infrastructure for years with its old circuit-switched infrastructure.

In the new network, switches read packet and cell headers. They refer to a directory that provides data type and encryption. More Web servers will be shifted to the network as well as hardware and software that oversee access rights. Other servers will be used for new applications like e-commerce. Users will specify whatever capacity they want. Provisioning is faster because switches can be configured remotely.

The new public net also eliminates overlays. Carriers will shift from existing data services to a single underlying infrastructure, eliminating duplicated resources.

Sonera is Finland's main long distance ATM backbone. It will initially feature three levels of data service. ATM is used by other carriers to carry voice and data on the backbone.

Sprint is planning the Integrated On-Demand Network (ION) which is an early part of the new public network. Calculated savings compared with its older equipment is 70% cheaper. Access lines are estimated to be 60% to 80%. Maintenance is estimated at 50% less.

In the new public network, other new products will allow users to plug in all their communications gear. Traffic from these sources is shunted to the carrier. These are more than superconcentrators. The technology delivers service classes based on bit rates like most ADSL (asymmetric digital subscriber line) services. AT&T is planning an equivalent service code-named INC (Integrated Network Connect). Other companies like Xedia offer Access Point which lets carriers offer bundled services with bandwidth management over a single access line. The new public network should allow carriers to cut costs and roll out new services more rapidly.

In the new infrastructure, virtual networks will not have much in common with their rudimentary equivalents where every connection must be manually configured. The new public networks will eliminate most configuration chores and run over managed backbones which guarantees throughput and latency for different types of traffic. Carriers will be able to add services on the fly.

Intelligence will also be added in the form of directory-enabled provisioning. Other products will become available that will deliver complete security so that even banks will do wire transfer over public networks.

References

Braun, Cedric R., "Wireless LAN Standard Proposal Highlights the Need for Speed," *Wireless Systems Design*, Vol. 3 Number 9, September 1998, pp. 13-16.

Feuerstein, Marty, and Michael Zhao, "Smart Antennas Tune in to CDMA Cellular Networks," *Wireless Systems Design*, Vol. 3 No. 6, June 1998, pp. 38-40.

Heywood, Peter, and Andrew Cray, "Building the Brave New World - Conventional Circuit Switching Won't Cut it in the 21st Century," *Data Communications*, Vol. 10 No. 20, October 1, 1998, pp. 44-49.

Mera, Narciso, and Robert T. Franzo, "Contrasting System-level for DSP-based Wireless Systems," *Wireless Systems Design*, Vol. 1 Number 9, November 1996, pp. 36-41.

Navarrete, Angela, "Web Warp Speed," *PC World*, Vol. 17 No. 1, March 1999, pp. 164, 170.

Peterschmidt, Eric, "Software Isn't Enough: Your State-of-the-Art Building Controls System," *Buildings*, Vol. 91 No. 4, April 1997, p. 30.

Schneiderman, Ron, "Competition Heats Up Over Third-Generation Concepts," *Wireless Systems Design*, Vol. 3 No. 1, January 1998, pp. 22-25.

Sweeney, John, "New DSP Architectures Put Desktop Capability in Pocket Communicators," *Wireless Systems Design*, September 1998, Vol. 3 No 9., pp. 37-42.

Taglione, John and Jeff Molson, "IR Provides High-speed Wireless Connectivity for Portable Devices," *Wireless Systems Design*, Vol. 2 No. 5, May 1997, pp. 45-46.

Index

O

P

Q

R

S

T

U

V

W

Z